ELECTRIC BICYCLES

ELECTRIC BICYCLES
A Guide to Design and Use

WILLIAM C. MORCHIN
HENRY OMAN

IEEE PRESS

A JOHN WILEY & SONS, INC., PUBLICATION

Library of Congress Cataloging-in-Publication Data is available.

ISBN-13 978-0-471-67419-1
ISBN-10 0-471-67419-2 (acid-free paper)

Printed in the United States of America.

10 9 8 7 6 5 4 3 2 1

TL
410
.M66
2006

CONTENTS

PREFACE

A battery-powered electric bicycle delivers personal travel at the lowest possible cost of propulsion energy. The traveler who pedals a bicycle derives muscle energy from food. If the food were cheese, then it would have to cost less than 15 cents a pound to compete with electricity costing 6 cents/kWh. However, an electric bicycle design is similar to an airplane design in that high efficiency is essential, and excess weight that has to be hauled over hills must be avoided.

This book describes guidelines and uses for battery-powered electric bicycles. It delivers to the reader the techniques, data, and examples that enable him or her to understand and predict the performance of a bicycle in terms of the speed and travel distance that a given propulsion configuration can deliver. The reader will be introduced to commercial versions of the electric bicycle through descriptions and tables of performance. Charts and data will permit this reader to understand performance and cost trades among the available commercial electric bicycles or make system engineering choices for a bicycle that he or she may decide to design, construct, or use.

The important components of an electric-powered bicycle are the battery, speed control, propulsion motor, and the speed reducer which couples the motor to the bicycle's wheel. The performances of electric bicycles that we built are a basis for the data we discuss.

America is heading toward a crisis that bicycling has solved in other nations. Growing urban areas require workers to travel ever-greater distances and spend hours each day commuting to work. Freeway expansion has been limited by the cost of acquiring rights-of-way and demolishing buildings. Interurban trains require alternative travel means for getting to stations and to workplaces. The cost of downtown parking already causes workers in Finland to bicycle to their Helsinki workplaces. Imported petroleum for fueling America's cars is getting more expensive. Furthermore, Americans generate more per-capita greenhouse gases than do the residents of any other country. The weather effects and ocean pollution effects of carbon dioxide emissions are becoming recognized, and America is being pressed to reduce its carbon dioxide releases.

In the rest of the world the quantity of bicycles being manufactured exceeds that of automobiles. Electric bicycle production is growing in the Orient and Europe. These bicycles, being powered with lead–acid and nickel–zinc batteries, are consequently heavy and limited in range. Now small high-performance lithium ion and nickel–metal hydride batteries are being manufactured and adapted for electric bicycle use. For example, a 3-kg (6.5-lb) lithium ion battery can propel an electric bicycle a distance of 40 km (25 miles) between charges, but in 2004 that battery would have cost of about $750. When such high-performance batteries become much less expensive, electric bicycles will become more common. Already electric bicycles powered with heavier lead–acid batteries are practical for traveling distances up to 25 km (15.5 mi) to work or to a commuting rapid-transit station. A 4.5-m (15-ft) wide freeway lane can deliver at best only 2000 automobiles an hour. The same width bicycle path can deliver 8000 bicyclists an hour.

Many factors must be considered when designing and constructing an electric bicycle. In 6 years of electric bicycle design, construction, and testing, we have learned how to optimize the bicycle design for meeting specific objectives. With our recorded design techniques and data, we are now able to help craftsmen who want to design or build their bicycles. We can also help consumers who are evaluating commercially available bicycles for their specific needs. Professional bicycle designers who seek optimum performance with specially designed components can also benefit from our experience and measured data that are not otherwise available.

We also give commentary and specific data on alternative choices for future modifications that improve performance. The purchaser of a factory-built electric bicycle will be able to understand the performance capabilities of alternative bicycle models before selecting the one to buy.

ACKNOWLEDGMENTS

We appreciate the support given in the development of electric bicycles and on the writing of this book by Mike Skehan. He is a builder *extraordinaire*, and we learned many things from him. With his help we were able to go from a requirement to build a demonstrator electric-powered bicycle to a finished product within 1 month of off-hour part-time effort. We also appreciate the encouragement given to us by another electric vehicle builder, Dave Cloud, a fellow member of the Seattle Electric Vehicle Association. Also appreciated is the special attention we received from Frank Jamerson in providing his two books and videos on the 1995 and 1996 Intercycle Conferences in Cologne Germany and Shanghai China, respectively. He also made available to us a hub motor that we used to demonstrate the ability to travel 130 miles from Auburn to the Oregon border south of Longview, Washington. Contact information for Frank Jamerson is given in Chapter 1 references. We also appreciate the friendly help given by Eric Sundin of EVs Northwest who permitted us visits to learn about the many available commercial electric bicycles and for greater exposure to various wheel hub motors and permission to test them.

We also acknowledge the time and guidance given to us by the following engineers: Floyd A. Wyczalek (President FWLilly Inc), Binod Kumar (Consultant, Infrastructure in Developing Nations), and Chris Johnson (Battery Research Specialist).

We are especially appreciative of the patience, help, and encouragement given by our wives, Earlene M. Oman and Karen E. Morchin, during the year that went into the preparation of this book.

WILLIAM C. MORCHIN
HENRY OMAN

Auburn, Washington
Normandy Park Washington
September 2005

.

CHAPTER 1

ELECTRIC BICYCLES—HISTORY, CHARACTERISTICS, AND USES

1.1 INTRODUCTION

We often hear the challenge: "Why would you want to use an electric-powered bicycle when the whole purpose of a bicycle is to get exercise?" Certainly this is true when everyone can so easily unlock the car door, get in, start an engine, and drive off. Where we live, you can on a warm sunny day go to a trail where you will see joggers, skaters, and bicyclists getting exercise. When you travel across the country on those highways that are not freeways, you'll see bands of cyclists on the open road with their panniers full of staples, supplies, and bedding. They are looking for an adventure.

Think back, was the bicycle invented for exercise? Probably not, for in the bicycle-invention days people got their exercise from walking to workplaces and working. Shortly after bicycles became available, the Western Union boys were on bicycles delivering telegrams. Today farmers in China haul live pigs to markets in panniers on bicycles. Workers in Helsinki, Finland, bicycle to their downtown offices to avoid high cost of auto-parking spaces. In the Netherlands 30 percent of personal travel is done on bicycles. We suggest that powering a bicycle with an electric motor has as much merit as powering an automobile with a gasoline engine. We still have fun with our automobiles, and we can now have fun with an electric-powered bicycle.

A traditional bicycle is a two-wheel vehicle that is propelled by the rider who delivers muscle power through pedals that rotate one of the vehicle's two wheels. The rider keeps the bicycle upright by steering the front wheel to create a force that restores the vehicle's center if gravity to its stable zone whenever necessary to prevent tipping. Today's motorcycle is a two-wheel vehicle that is propelled by a fuel-burning engine. An electric bicycle carries batteries or fuel cells that deliver electric power to a motor that is coupled to either wheel. In most electric bicycles the rider can chose to use muscle power to deliver all, part, or none of the propulsion power required to maintain his or her adopted travel speed. Some models even sense your pedal pressure and command the motor to deliver more power whenever you pedal hard.

Many electric bicycles are specifically designed and built for travel. Average travel speed, when compared to pedaled-only bicycles, can be increased by 8 to

Electric Bicycles: A Guide to Design and Use, by William C. Morchin and Henry Oman
Copyright © 2006 The Institute of Electrical and Electronics Engineers, Inc.

10 km/h (5 to 6 mph) above the speed an average person could travel by pedaling. An exception is the athlete or world-class competitor. These athletes can pedal a bicycle faster than an ordinary cyclist can move on a conventional electric bicycle.

Our objectives in this book are as follows:

1. Describe the performance of presently available electric bicycles and the performance that they can achieve.

2. Quantify the design factors that affect electric bicycle performance in terms of vehicle weight, speeds on road grades, and travel distance available between battery recharges or fuel cell refueling.

3. Predict coming improvements in weight and life of energy storage technologies, propulsion efficiency, and use of electric bicycles.

4. Show how application of today's systems-engineering techniques can produce electric bicycle designs that have the long travel range and low life-cycle cost that make them more applicable to the world's travel needs.

1.2 HISTORY OF BICYCLES

Pierre Lallenent in France first conceived the two-wheel pedal-powered bicycle in 1862 and demonstrated it the next year. By 1870 wire spokes had been developed, and nipples for tightening spokes came in 1874. In 1871 James Starley, a 67-year-old foreman at Coventry Sewing Machines, patented the Ariel bicycle, which as an option offered levers that doubled the distance traveled per pedal stroke. In 1880, when the League of American Wheelmen was formed, the "ordinary" bicycle had a small rear wheel and a front wheel that had a diameter of up to 5 ft. Pedals were solidly coupled to the front wheel. The average bicycle weighed 50 lb, but some were as light as 21 lb. In "century runs" athletic bicyclists on "ordinaries" pedaled 100 miles in 1 day during an age in which the self-powered alternative was hiking, with a practical limit of 20 miles.

Chain drives and pneumatic tires, invented by John Dunlop in 1888, made possible the *safety bicycle* in which the two wheels had the same diameter. Now the rider could at any time touch the ground with his feet, and crank the pedals at an optimized speed that differed from the wheel speed. Pneumatic tires made it possible to ride over rough roads with reasonable comfort.

The Sturmey Archer rear-wheel hub, which contained an epicycle gear train, also called *planetary gears*, was patented in the 1901 to 1906 period. With the first hubs the cyclist could select a "lo" gear ratio when climbing hills, and a "hi" ratio for fast bicycling on level ground with a tail wind. Later designs had three ratios. In the high-speed position the rear wheel was coupled directly to its driving sprocket. The latest version achieved five ratios of pedal speed to wheel speed with two planetary gear sets.

Sturmey Archer hubs lost popularity when DeRailleur gear shifting became practical. On DeRailleur-equipped bicycles, the cyclist can shift the chain among up to seven "free-wheel" sprockets that have various diameters. These sprockets are coupled to the rear wheel through an over-running clutch, which permits coasting downhill without the pedals being turned by the chain. The cyclist can also select one of three "chain wheel" sprockets that have different diameters. The chain wheel assembly is coupled solidly to the pedal cranks. With the now-available gear ratios, the bicyclist can crank the pedals at a pedal speed that delivers maximum muscle power. The bicycle then moves at a speed in which all net power is consumed in the air drag, friction, and hill climbing.

1.3 HISTORY OF ELECTRIC BICYCLES

Electric bicycles first appeared in the late 1890s. On December 31, 1895 Ogden Bolton, Jr., was granted U.S. Patent 552271 for a battery-powered bicycle with a 6-pole brush-and-commutator direct-current (dc) hub motor mounted in the rear wheel (Fig. 1.1). The motor contained no internal gears. It could draw 100 amperes (A) from a 10-V battery.

On December 28, 1897, Hosea W. Libbey of Boston was issued U.S. Patent 596272, for an electric bicycle that was propelled by a "double electric motor" that was in the hub of the crankshaft axle (Fig. 1.2). The 5-pole brush-and-commutator dc motor drove two closely spaced rear wheels, a configuration that Libbey had invented earlier in 1893. His patent showed crank rods for carrying the motor torque to the rear wheels. However, he soon adopted the sprocket wheel and chain drive for delivering power to the rear wheels. He also used a double battery, consisting of cylindrical cells with a central partition, which he invented in April 1895. One-half of the battery was to be used for travel on level ground, and both halves were used when climbing hills. The battery acid was refilled from a reservoir mounted under the rider's seat. A fabric saturated with the diluted sulfuric acid separated the battery plates. This avoided the spilling of sulfuric acid.

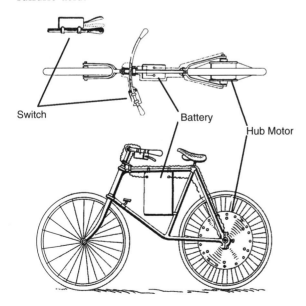

Switch   Battery   Hub Motor

Figure 1.1   First electric bicycle, invented by Ogden Bolton, Jr., in 1895, was propelled by a gearless hub motor in the rear wheel and had a 10-V battery.

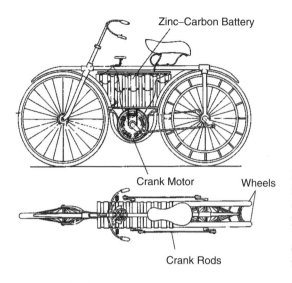

Figure 1.2 Hosea Libbey's invention, U.S. Patent 596,272, of the "crank-axle hub-motor powered electric bicycle," in 1897, featured crank rods that couple the twin rear wheels to the motor.

In 1898 in Chicago, Illinois, Mathew J. Steffens invented an electric bicycle on which the driving belt rode on the periphery of the real wheel (U.S. Patent 613752). The rounded belt rode in a tire groove and eliminated belt slippage problems by applying force at its contact with the ground (Fig. 1.3). The motor, mounted on the seat post, drove the belt pulley through sprocket wheels and sprocket chain. Steffens said his idea could be applied to "tricycles and other similar vehicles."

John Schnepf in New York patented a design in which a pulley on the motor shaft delivers propulsion power to the top surface of the rear wheel of the bicycle, as shown in Figure 1.4. The U.S. Patent number is 627066. He also suggested that the battery could be charged by the motor acting as a dynamo when the bicycle coasts downhill. However, he admitted that the battery would also need the normal means for fully recharging it after a trip.

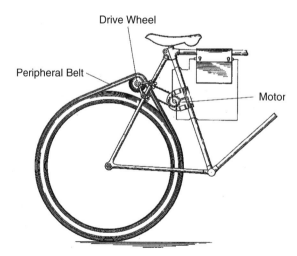

Figure 1.3 Mathew J. Steffen's invention, U.S. Patent 613,732, of the "wheel periphery belt drive electric bicycle," in 1898. The outer surface of the driving belt delivers propulsion power at the road surface.

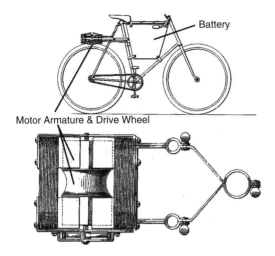

Figure 1.4   John Schnepf's invention, U.S. Patent 627,066, in 1899, is the "friction roller-wheel drive electric bicycle."

In 1969 G. A. Wood, Jr., expanded on the friction-wheel drive in his U.S. Patent 3,431,994. He used multiple subfractional horsepower motors, each rated less than $\frac{1}{2}$ horsepower. Four motors were coupled together to drive the bicycle's front wheel through a friction drive wheel that pushed against the bicycle's front wheel. Each motor powered the friction drive through a set of series-connected gears, as shown in Figure 1.5. Torque sensing and control of the power supplied by the motor was developed in the late 1990s. Takada Yutka of Suwa, Japan, filed a patent in 1997 for such a system. Shu-Shian of Taipei and others followed him in 2001.

In the 1920s a rear bicycle wheel with an integral gasoline engine was offered. Also, a small handle-bar-supported engine was sold for driving the front wheel with a tire-riding roller. Larger engines were mounted between the rider's legs on vehicles called "motorcycles."

Hugo Gernsback proposed electric bicycle propulsion in his January, 1924, *Science and Invention* magazine. His concept featured an underground cable that radiated radio-frequency (RF) power, collected by an antenna on the bicycle.

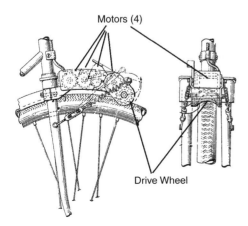

Figure 1.5   G.A. Wood, Jr.'s, invention, U.S. Patent 3,431,994, in 1969 for "multiple use of subfractional horsepower motors to drive an electric bicycle."

In 1992 Vector Services Limited offered a practical electric bicycle, the Zike. A nickel–cadmium battery built into a frame member and an 850-g permanent-magnet motor powered it. The Zike is 115 cm (45.3 inches) long and 100 cm (39.4 inches) high. It weighs 24 lb and can carry a rider weighing "up to 17 stones." With the Zike doing all the work, the range is "up to one-hour's riding." Having Zike doing just most of the work extends the range by a half-hour. Pedaling normally, with Zike helping on hills and during strong headwinds, gives 3 hours of travel on one battery charge. Other technology for improving the next electric bicycles was proposed in [1].

1.4   SOME USES FOR THE ELECTRIC-POWERED BICYCLE

Possible uses for electric bicycles include recreation, commuting to work, delivery of goods and services, and establishing communication with remote villages in developing nations.

1.4.1   Recreation

Electric propulsion can increase the distance that a physically weak person can travel on a bicycle. However, most recreational bicyclists in America are motivated by (1) the desire to maintain the cyclist's physical condition and (2) the challenge of achieving an unusual goal. For example, each year bicycle clubs in Washington and Oregon sponsor a 193-mile Seattle-to-Portland 2-day ride. This challenge typically attracts some 10,000 cyclists in early summer each year.

1.4.2   Commuting to Work

In traffic jams on freeways overloaded with commuters, the slowly moving cars create local air pollution. "Light rail," is a proposed solution. However, it offers the same rider inconvenience features that bankrupted the interurban trains in the 1930s. Driving straight from one's garage to one's workplace is simply more pleasant than the alternative of walking in the rain to a depot, riding a train that stops every mile or so to pick up passengers, and then transferring to a bus for the final leg of the journey.

Portland, Oregon's, $214 million MAXI commuting rail expansion illustrates the point. The environmental impact statement forecast that riders would grow from 24,000 per day to 42,500 by 1988. Riders actually dropped to 18,000 by 1988. Meanwhile, Portland area's daily freeway use grew to 4 million vehicles per day. Apparently each voter who approved MAXI bonds expected the freeways to be cleared for his or her own driving [2].

The high pollutant content of air in downtown zones of cities will not be cured by imposing more and tighter emission controls on our out-of-town coal-burning and oil-burning steam-power plants. The effective cure will be limiting the number of automobile engines in downtown zones, either with barriers or high parking fees. Then the commuters' options are riding the public transportation or bicycling. In cities in Europe and Asia the crowded bicycle lanes demonstrate that the bicycle can be the better choice.

In commuting, an electric bicycle could be useful. By increasing travel speed it can make bicycle lanes more productive. For example, one freeway lane, when converted to bicycle use, can deliver 8000 commuters per hour. This lane can deliver only 2000 automobiles per hour if they travel at an optimum speed of 35 miles (56 km) per hour [3].

1.4.3   Delivery of Goods and Services

In Germany, because of aging postal workers, the postal employers have concluded that electric bicycles will help get the mail delivered faster and easier. In Shanghai one pizza company delivers its product with the aid of electric bicycles.

1.4.4   Police and Army Applications

Police on bicycles can silently and quickly patrol public places in urban areas. Electric-propelled bicycles add to the quickness and endurance of the policeman. Police in cruising patrol cars leave court summons with indicated fines on cars that are parked in spaces where parking time has expired. This is not an efficient procedure because the policeman has to find a place to park his patrol car in a packed downtown zone and walk to the offender's site. Electric bicycles solve this problem. The policeman can lean the bicycle on the post that supports the meter and write the summons for the owner of the illegally parked car. Also an agent who rides on an electric bicycle can quickly collect deposited coins from the meters.

1.4.5   Opening Communication to Remote Areas

In developing nations, an important immediate use of electric bicycles can be opening communication to remote villages for establishing educational opportunities. Promising students could then commute to schools.

Replacing parts that have failed, worn out, or broken in an accident can be costly in a community with only a few dozen vehicles. The electric bicycle can alleviate the lack of sufficient number of vehicles. If adequate roads for automobiles are not available, the electric bicycle can make the trip on trails more endurable.

1.5   EXAMPLES OF ELECTRIC BICYCLES

Examples of commercially available electric-powered bicycles are shown in Figure 1.6. The first three photographs show (a) the LA Freezzz made by Giant Bicycle (United States), (b) the Power Cycle model 550 made by Mirada Bikes (Taiwan), and (c) the Enviro made by EV Global (United States). Bicycles in Figure 1.6a, 1.6b, and 1.6c illustrate what is referred to commonly as the obvious "step-through" design. These vehicles have frame designs that have been developed specifically for use as electric-powered bicycles. A more conventional frame style, but specifically made for electric bicycle use, is shown in Figure 1.6*d*. It is the Power Cycle model 500.

Three of the bicycles, (Fig. 6.1*a*, 6.1*b*, and 6.1*d*), use a motor mounted near the pedal crankshaft. The motor delivers power to the crankshaft through a gearbox and a free-wheeling clutch so that power can be shared between the motor and the force that the rider applies on the crank pedals. Each of these models delivers power to the rear wheel through an internal four-speed hub in the rear wheel. Each has a battery mounted either behind or in front of the post that supports the rider.

(*a*)

(*b*)

Figure 1.6   Four commercially available electric bicycles: (*a*) LA Freezzz, (*b*) Power Cycle model 306, (*c*) Enuice, and (*d*) Power Cycle model 500.

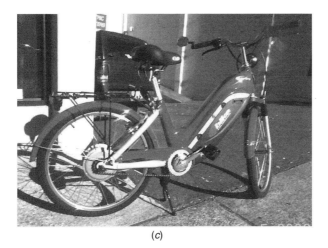

(c)

(d)

Figure 1.6 *continued.*

The remaining model, in Figure 1.6c, has a hub motor that is inside the center of the rear wheel. Thus the motor delivers power directly to the wheel. The battery is mounted on the slanted monotube frame, which is covered with a molded plastic fairing.

A closeup photograph of the hub motor is shown in Figure 1.7. The hub motor has a torque reaction arm that is fastened between the bicycle frame and the motor axle. Also, rather than the motor armature rotating, the motor field rotates. In a practical sense, the bicycle wheel rotates at a relatively slow rate, so internal gearing is required to match the optimum motor speed to the wheel speed. Motors could be designed to rotate at low rates. However, the efficiency of a motor rotating at low rate is much lower than that of a high-speed motor that delivers the same power output. Also, the slow-speed motor would weigh much more than the high-speed motor with its speed reduction gearing. These bicycles are further described in Table 1.1.

TABLE 1.1 Performance of Electric Bicycles Being Manufactured in Year 2004 [a]

Maker and Name	Design	Motor Power W	Travel Distance km	Maximum Speed km/h	Battery Voltage V	Battery Capacity Ah	Battery Energy Wh	Battery Type	Remarks
Electric Bikes (United States)									
Currie E-Ride		250	32	32	24	17	408	L-A	20 and 16-inch models
Currie E-Folder	E-Bike	250	32	32	24	17	408	L-A	
ETC Ne Century	E-Bike	250	21	21	24	12	288	L-A	
ETC Traveler	E-Bike	250	19	21	24	12	288	L-A	
EV Global Folding Mini E-Bike		400	19	24	36	7	252	Li Polymer	Standard option
EV Global Folding Mini E-Bike		400	39	24	36	14	504	Li Polymer	Extended range option
EV Global LE	E-Bike	500	32	29	36	8	288	L-A	
EV Global Police PE	E-Bike	500	32	29	36	8	288	L-A	
Optibike 400	E-Bike (1)	400	na	32	36	12	432	NiMH	34 mph with gearing
Rabbit Tool USA Drive	Pedelec/E-Bike (2)	500	32	32	24	8	192	NiMH	12 Ah option
Electric Bike Systems 3-Wheeler	E-Bike	200	48	23	48	7	336	L-A	Taiwan
Giant Lite	Pedelec	240	37	29	24	6.5	156	NiMH	
Panasonic Electric Hybrid Folder	Pedelec	240	29	24	24	3.5	84	NiCd	
Panasonic Electric Hybrid Folder				24	24	2.6	62	NiMH	
Aprilia Enjoy	Pedelec	250	40	25	24	13	312	NiMH	Italy

Name	Type								Country
Merida PowerCycle	Pedelec	230	40	24	24	9	216	NiMH	Taiwan
ZVO Power X	Drive on Dahon	400	32	32	24	7	168	NiMH	
Electric Bikes (Canada)									
Procycle Mikado	Pedelec	250	20	24	24	5	120	Ni-Cd	(5)
Energy & Propulsion Systems	Pedelec	250	24	32	24	8	192	NiMH	(6)
Electric Bikes (Europe)									
Prima Joe Fly	Pedelec	250	25	24	36	5	180	L-A	Cyclon Italy
Prima Runner	Pedelec	250	25	24	36	5	180	L-A	Cyclon Italy
Prima Yello Dream	Pedelec (3)	250	25	24	36	5	180	L-A	Cyclon Italy
Aprilia Enjoy	Pedelec	250	40	25	24	13	312	NiMH	Italy
Bikit	Folding Pedelec	100	32	18	24	7	168	L-A	Israel
Elektromobilbau Schachner	Pedelec/E-Bike	400	40	24	36	5	180	NiCd	Austria
Sparta Pharos	Pedelec (4)	250	40	24	24	7	168	L-A	Holland
Velocity Dolphin Blackpowder	Pedelec	270	20	30	24	7	168	NiCd	Switzerland
Heinzmann Estelle Comfort	Pedelec/E-Bike	400	27	21	36	5	180	NiCd	Germany
Piaggio Albatross	Pedelec	250	30	24	36	5	180	L-A	Italy
Yoker City	Pedelec	240	25	24	24	7	168	L-A	Germany
Electric Bikes (China)									
Tianjin Flying Pigeon	E-Bike	180	40	20	36	12	432	L-A	
Wuhan Yunhe	E-Bike	150	50	20	24	22	528	L-A	
Toprun Little Princess	Pedelec	180	60	28	36	5	180	NiCd	
Sunpex Swan	E-Bike	150	32	22	24	12	288	L-A	
SUEL Jet Bike	Pedelec	150	30	20	24	7	168	NiMH	
Changjie Electric Bicycle TD H09Z	E-Bike	160	50	20	36	6	216	L-A	

(continued)

TABLE 1.1 Performance of Electric Bicycles Being Manufactured in Year 2004[a] continued

Maker and Name	Design	Motor Power W	Travel Distance km	Maximum Speed km/h	Battery Voltage V	Battery Capacity Ah	Battery Energy Wh	Battery Type	Remarks
Electric Bikes (United States)									
Changjie Electric Bicycle TD H10Z	Folding E-Bike	160	40	20	24	12	288	L-A	
Suzhou Small Antelope	E-Bike	180	40	20	36	12	432	L-A	
T & Dl Continent Dove	E-Bike	180	55	20	36	12	432	L-A	
T & Dl Gentle Breeze	E-Bike	180	45	20	36	12	432	L-A	
Zhejiang Wolong	E-Bike	150	50	20	12	12	144	L-A	
Hengbo King Ring	E-Bike	250	50	20	36	7	252	L-A	
Shanghai Forever Career	E-Bike	180	50	20	36	12	432	L-A	
Xingyueshen LanFeng	E-Bike	150	45	20	36	12	432	L-A	
Hangzhou Qianjiang-Gear Gold Algret	E-Bike	180	50	20	36	12	432	L-A	
Jingcheng TDN32Z	E-Bike	200	60	80	36	12	432	L-A	
Hangzhou Dongking QunFang	E-Bike	180	50	20	36	12	432	L-A	
Electric Bikes (Japan)									
Miyata GoodluckAlumni 26	Pedelec	235	20	15	24	3.6	86	NiCd	
National Elegant ViVi-L	Pedelec	240	23	15	24	2.8	67	NiMH	

Bridgestone Assista Superlight	Pedelec	235	25	15	24	3.6	86	NiCd
Yamaha PAS Super Light	Pedelec	235	20	15	24	3.6	86	NiCd
Honda Step Combo		235	30	24	24	3.6	86	NiMH
Suzuki Love SNV 24	Pedelec	240	35	15	24	2.8	67	NiMH
Sanyo Enacle Leger	Pedelec	250	36	15	36	2.4	86	NiCd
Mitsubishi Bicic	Pedelec	235	41	24	24	4.5	108	NiCd
Maruishi Frackers Como Assist		235	25	15	24	3.6	86	NiCd
Electric Bikes (Tawian)								
Merida Powercycle	Pedelec	230	40	24	24	9	216	NiMH
Giant LaFree Twist	Pedelec	240	37	29	24	6.5	156	NiMH	Taiwan
Soleus SEB-26S	E-Bike	250	24	35	24	12	288	L-A
Eleon GP-26MTB	E-Bike	300	30	25	24	17	408	L-A
Cosmos Ocean EZ-1	E-Bike	250	20	24	36	4.4	158	NiCd
Averages =		251	35	24	29	8.49	246	

[a](1) Dual-suspension Monocoque aluminum frame, (2) on Dahon "Helius" folder, (3) carbon fiber frame and wheels, (4) Yamaha powered, (5) with 60 and 100 percent power assist modes, (6) to 40 m with 25, 50, 100, and 200 percent power assist modes.

Figures 1.8 and 1.9 show more examples of commercial electric bicycles. Figure 1.8 shows a motor and battery mounted on a more conventional bicycle frame design by ZAP (United States) on its model DX. Its two motors drive the bicycle wheel through a knurled knob that presses on the rear wheel. The rider must pedal to activate the motors. Some models use a single motor. Figure 1.9 shows a kit one can purchase to attach to the bicycle. The ZAP system is being adopted by some police bike forces. A folding bicycle frame model is shown in Figure 1.10.

Electric propulsion is available even in recumbent bicycles, such as the BikeElectric. We have made a three-wheel recumbent, with two rear wheels and one smaller front wheel, and covered the frame with a plastic fabric. We drove the left rear wheel with a chain through a Sturmey Archer gear shifter and drove

Figure 1.7   Propulsion motor inside the hub has a rotating field that is coupled to the hub by a speed-reducing gear. A torque reaction arm couples the motor armature to the bicycle frame.

Figure 1.8   The DX kit offered by ZAP of USA, is available to an owner who decided to add electric propulsion to his bicycle. The battery-powered motor's output shaft pushes the surface of the rear tire. The battery is in the "ZAP" bag. (By permission from ZAP, Santa Rosa, CA, Zapworld.com).

Figure 1.9   The ZAP SX-bike kit, offered by ZAP of USA, is the yellow box that contains a headlight, battery, and a motor with an output shaft that propels the outer surface of the bicycle's front tire.

Figure 1.10   Hub motor in its rear wheel propels the ZAP World.COM foldable electric bicycle that is designed for travelers.

the right rear wheel by pedals through another Sturmey Archer gear shifter. We established a record for low rate of energy consumption in the Seattle-to-Portland Bicycle Classic in 1996.

The most pertinent characteristics of some recently available commercial electric-powered bicycles are summarized in Table 4 that was compiled by Jamerson [4]. When we started our research on electric-powered bicycles in late 1992, there were virtually no commercial electric bicycles available. By early 1998 there were at least 49, as shown by Jamerson [5], [6]. Gardner [7] observed that although bicycle production has been decreasing from its peak at 107 million units a year in 1995, the production of electric bicycles is increasing. That data and those from Ed Benjamin [8] show that electric bicycle production has been growing and was expected to grow at a rate of about 35 percent per year over the 11-year period from 1993 to 2004. Benjamin reported that there were

160 manufacturing exhibitors at the 2003 Shanghai Cycle Show. His predictions are plotted in Figure 1.11, which shows that the People's Republic of China dominates in electric bicycle production. Recent data show that China's bicycle production had grown to 1.6 million a year, by 2002.

We also noted some distinctions in reviewing the performance of electric bicycles being produced in the world, as reported in Table 1.1 of Benjamin [8]. One is that when NiMH and NiCd batteries are used in place of lead–acid batteries, battery capacity is reduced. The main incentive for this was to avoid changing the vehicle cost. Thus, travel distance was sacrificed to avoid raising the vehicle's selling price. Second, we noticed that mainly the U.S. manufacturers offer more kits rather than complete bicycle assemblies. The reasons could be savings in cost, limiting the manufacturer's liability exposure, or relative lack of interest in electric-powered bicycles in the United States.

Another distinction is that the upper speed of U.S. bicycles generally ranges from 25 to 35 km/h, whereas in other countries it ranges from 15 to 25 km/h. Also bicycles made in China generally have a price below the worldwide average price. Another distinction for the user is in Figure 1.12, which shows the manufacturers'

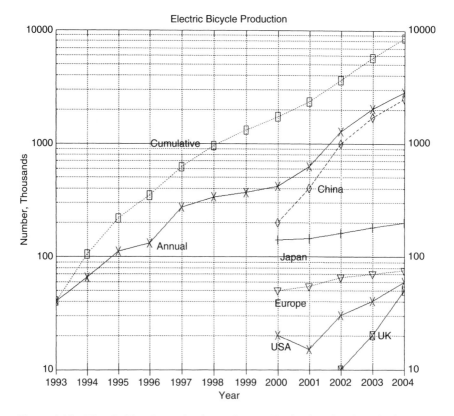

Figure 1.11   Electric bicycle production estimates. Benjamin sales data [8] for years 2000 through 2004 are merged with Gardner [7] production data for years 1993 through 2000.

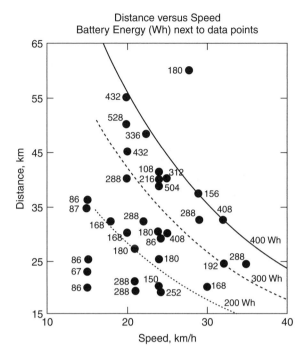

Figure 1.12   Travel distance versus speed for representative manufactured electric bicycles. Battery energy capacity values in watt-hours are placed next to data points when possible. The graph lines show theoretical computed values assuming 100 percent system efficiency and 100 percent depth of discharge for batteries having the indicated energy capacities in watt-hours.

stated distance capability as a function of travel speed. Next to the data points is shown the energy capacity, in watt-hours (Wh), of the battery pack for the corresponding bicycle. This battery capacity is the product of the nominal battery voltage and ampere-hour rating. Overlaid onto the data points are a set of parametric curves, for the theoretical performance, assuming three values of energy, in watt-hours. We found that electric bicycle performance was often overstated when comparing the data points with the curves. This is particularly true for the data point shown for a speed of about 28 km/h and a distance of 60 km using energy of 180 Wh. This electric bicycle made in China is one where the motor does not run unless the rider pedals (a Pedelec). Undoubtedly, most of the distance is covered by pedaling. The curves were calculated using the procedures, which are explained in Chapter 2, for level travel at sea-level altitude for a 75-kg (165-lb) rider on a 36-kg (80-lb) bicycle. The curves were based on a system efficiency of 100 percent. We have found that system efficiency is generally in the 55 to 65 percent range.

Some interesting averages can be calculated from the data in Table 1.1. For example, the average motor power is 250-W, the average battery capacity is 8.5 Ah, and the corresponding energy stored in the battery is about 250 Wh. Average speed is 24 km/h and the travel distance is 34 km.

1.5.1   Electric Scooters

A close cousin to the electric-powered bicycle is the electric-powered scooter, which is a two-wheel vehicle that has no means of human-powered propulsion.

It has a platform on which the rider either stands in some designs or sits in other designs. The wheels are usually smaller in diameter than those on a bicycle. Scooter sizes range from those that can be carried with one hand to others that are too heavy to lift by an average person. The smaller electric scooters travel at electric bicycle speeds. The larger ones travel at city-street speed limits.

The world production of small electric scooters has been an order of magnitude less than that of electric bicycles. However, scooter sales are growing at a rate of about 47 percent per year. This is much higher than the growth rate of electric bicycle sales. In the United States the small electric scooter has a greater popularity than the electric bicycle has.

A new product is the Segway, an electromechanically stabilized scooter on which the rider stands erect on a platform between two wheels (Fig. 1.13). Each wheel is propelled by an electric motor. The scooter maintains a level platform as it passes over terrain-slope variations. The rider controls stopping and starting functions by leaning forward or backward, or twisting upright handle grips in the appropriate direction. It is designed to be used on pedestrian sidewalks. It has a handgrip for adjusting its travel speed.

The Segway is a relatively expensive machine with perhaps an equal number of negative and positive opinions of its utility and safety. Simplicity is not its

(a)   (b)

Figure 1.13 Segway, an electromechanically stabilized scooter: (a) from U.S. Patent 6,367,817 B, Kamen et al., April 9, 2002 and (b) from U.S. Patent 6,543,564 B1, Kamen et al., April 8, 2003.

forte. The design philosophy of using two side-by-side wheels requires a complex stabilization system. In contrast, a bicycle with two tandem wheels is easy to balance while traveling. The Segway weighs 80 lb (36 kg) and its top speed is 12 mph (19 km/h) and has a travel distance capability of 11 miles (18 km). An analysis of its performance in a hill-climbing test is given in Chapter 2.

1.6   FUTURE OF ELECTRIC BICYCLES

Economics, the cost of gasoline and diesel fuel, and air pollution will affect the use of electric bicycles and bicycles in general. Today's fuel price in some nations is related to the cost of production and in other nations to the need to control imports. The price of fuel in the future will be affected by the exhaustion of reserves of petroleum and natural gas, and the need to limit (1) the pollution of the environment and (2) the generation of carbon dioxide.

Atmospheric pollution by motor vehicles in cities can be prevented only by prohibiting these vehicles in downtown zones. One type of bus, used in Seattle, is propelled by a diesel engine when traveling in the suburbs and by electric-trolley power in city-center tunnels. Buses and light rail suffer from the rarely mentioned user-inconvenience features, which drove the electrified interurban railroads into bankruptcy in the 1930s. Again bicycles, and especially battery-powered electric ones, can become common vehicles for commuting to downtown workplaces.

The problems in bringing every day a million people to work in a skyscraper-filled downtown of a large city are easy to identify. New York City built a useful subway network with immigrant labor at a time when environmental questions did not delay construction. Today subway riders grumble but chose not to go to work in buses or cars. A subway network feeding a downtown zone would not work in Los Angeles or Seattle where the big employers are in outlying suburbs.

We can look to European and Chinese cities to see that bicycling to work-places is a solution that worked. Every day in Shanghai over a million people bicycle to work. Availability of electric bicycles would further enhance the bicycle for commuting to work.

A freeway lane built for automobiles could easily be converted to a bicycle lane, quadrupling its commuter-delivering capacity. The alternative of quadrupling the number of automobile lanes would require a hopeless environmental impasse. The pavement and bridges would need to carry only riders on bicycles that weigh less than 300 lb each, rather than multiton cars and buses. Bicycling to work could be made more comfortable with rain covers over the bicycle ways (see Chapter 8). The supporting infrastructure could include coin-operated recharging stations at workplaces.

Commuter trains that stop at frequent intervals result in slow travel from a suburban home to a downtown workplace. One solution is to have greater separation between stations and parking garages for local commuters who drive from home to the station. At one such station in Virginia, by 9:00 A.M. on a weekday the parking garage is full, and every nearby street-side parking spot is occupied. Electric bicycles for home-to-station travel would solve this problem. A lightweight foldable electric bicycle that the rider could carry on the train and bicycle from the downtown station to his workplace would be even better.

In this new environment electric bicycle performance will improve in terms of travel range on a charged battery, battery life, reliability, riding comfort, and many other features. New electric bicycle builders will spend substantial money on improved bicycle performance in order to capture a leading marketing position from the existing producers. For example, China will undoubtedly market low-cost electric bicycles from a high-production factory. In our American consumer market, a new better performing electric bicycle can become popular for a limited time, even if it costs more.

We expect to see significant and continuing improvements in the travel range of electric bicycles from improvements in the energy content of batteries. One example is the zinc–air fuel cell, which is being developed for electric cars. It can't be recharged by the owner but could be recharged at "filling stations" where a discharged fuel cell can be quickly exchanged for a recharged fuel cell. Hydrogen storage in nanotube grids is another coming technology.

1.7 LAWS AND REGULATIONS GOVERNING ELECTRIC BICYCLES

Regulations that will affect the design and use of electric bicycles are evolving throughout the world. The approach is to define the electric-powered bicycle as a class of bicycle and then impose an upper limit on bicycle speed when motor powered. This limit is then related to speeds normally observed with regular bicycles in the city. The regulations that define design limitations are summarized in Table 1.2.

Other regulations deal with insurance, helmets, operator age, and travel on roads and pathways. Whenever existing limitations are exceeded, the regulations are made stricter in terms of application. Motor-vehicle-related regulations cover topics such as licensing, road use, required insurance, and age limits of drivers. They may be applied to electric bicycle riders as soon as their presence on streets becomes significant.

In the United States electric-powered bicycles are consumer products, as defined by the federal government. The City Electric International Consulting Group cites possible regulations in Table 2 of [9]. The electric bicycle has two or three wheels and is powered either by pedals or an electric motor. Likely regulations would restrict the motor power to a value of less than 750 W. The bicycles' maximum speed on a paved level surface might be held to less than 20 mph when the rider is a person weighing over 170 lb, and the bicycle is powered only by the motor.

Very specific standards limit the voltage of conductors to which a vehicle-occupant might be exposed. For example, engine-starting batteries in automobiles are generally limited to 40 V for safety reasons.

TABLE 1.2   Summary of Regulation Limitations for Defining an Electric Bicycle in the Bicycle Classification

Location	Motor Power Limit (W)	Speed Limit (km/h)	Weight Limit (kg)	Other Limitations
European Union	250	25 [a]	No	Pedal assist [a]
United Kingdom	200 [b]	25 [c]	40 [d]	Must have pedals and on/off switch
Canada	500 [e]	32	No	Must have pedals and less than 4 wheels and on/off switch
Taiwan	No [f]	30	40	
Japan	No [g]	24	No	
China	240	20	No	Must have pedals
United States	750	32.2 [h]	No	Must have pedals and less than 4 wheels

[a] Motor power is progressively reduced as speed increases to 25 km/h at which speed motor power must be zero.
[b] 250 W for tricycles.
[c] Maximum with motor power applied.
[d] 60 kg for tricycles.
[e] Motor power zero for speeds less than 3 km/h.
[f] Power only engages during pedaling.
[g] Motor power equals pedal power up to 20 km/h, thereafter progressively reducing to zero at 24 km/h.
[h] When the operator weighs 170 lb or less.
Source: From Jamerson [4].

In the United States there will also be the pertinent state laws, rules, and regulations with which to be concerned. Each state will define where the bikes can go, how fast they can go, and how the state's insurance codes apply. Some states may require inspection and certification of electric bicycles. Firms that manufacture, sell, or service electric bicycles may be required to have permits or licenses. Some states might define an electric bicycle as a motorcycle. Then, in order to obtain a driver license, the electric bicycle rider would have to pass a special written test and also a driving test. Helmet requirements may be the same as those for the conventional bicycles. The electric bicycle would normally not require registration, licensing, or operator qualification.

Some other nations regulate the application of electric motor power to bicycles. For example, some require that motor power alone must not propel the bicycle. Some countries say that the motor power must not exceed the operator pedal power, while others say the pedals must be used before motor power is applied. Many have some speed limit at which the motor power must be cut off.

1.8   CONCLUSION

In this chapter we summarized the achieved performance, history, where it is made, who makes it, the costs, the uses, and the future of the electric bicycle. The following chapters cover the basics of electric propulsion, the batteries, motors and controllers, and the system engineering principles for designing a successful electric bicycle, testing of components, and forecasts of some of the future developments that we foresee.

REFERENCES

1. James B. Coate, Design of an Efficient and Economical Electric Bicycle, Masters of Science in Engineering Design, Tufts University, May, 1994.
2. G. Scott Rutherford, Light Rail, Heavy Politics, *Pacific Northwest Executive*, 1989, pp. 18–22.
3. Michael A. Brown, Electric Bicycle Transportation System, Proceedings of the IEEE-sponsored 37th Intersociety Energy Conversion Engineering Conference, 2002.
4. Frank E. Jamerson, *Electric Bikes Worldwide 2002: With Electric Scooters & Neighborhood EVs*, January 15, 2002, 6th ed., Electric Battery Bicycle Company. Frank Jamerson, Electric Battery Bicycle Co., Naples, FL and Petoskey, MI, Publisher, can be contacted at e-mail elecbike@aol.com or www.EBWR.com<http://www.EBWR.com>.
5. Frank E. Jamerson, Electric Bikes Worldwide: Intercycle Cologne 95, Electric Battery Bicycle Co., Naples, FL and Petoskey, MI, September 14–17 1995.
6. Frank E. Jamerson, Electric Bikes Worldwide 97: China Exhibition, Shanghai/Interbike Anaheim, Electric Battery Bicycle Co., Naples, FL and Petoskey, MI, January 20, 1997, 3rd ed.
7. Gary Gardner, Bicycle Production Down Again, *Vital Signs 1998*, World Watch Institute, p. 89.
8. Ed Benjamin, Cycle Electric International Consulting Group, Nov. 12, 2003, website: www.cycleelectric.com.
9. City Electric International Consulting Group, Newsletter, November 2002.

CHAPTER 2

FUNDAMENTALS OF ELECTRIC PROPULSION

2.1 INTRODUCTION

Much technical ground is covered in this chapter. The equations and tables provide data with which you can design your electric bicycle. If you plan to buy one, there is enough information here for you to evaluate the validity of the performance claims that the vendor makes. For your purchase decisions this information will enable you to select the best performing electric bicycle.

Developed in the first section of this chapter is a mathematical model that predicts the power consumed in bicycle travel. The power needed for hill climbing, overcoming wind resistance, and rolling resistance is developed in analytic form. Pertinent data include the coefficient of aerodynamic drag of shapes of interest and the projected area of the bicycle rider. The bicycle's rolling resistance is related to total weight being carried by its wheels.

We conclude the chapter with a discussion of life-cycle cost and illustrate our analytical procedure with two interesting vehicles: a pedaled racer and a standup electromechanically stabilized scooter. Chapters 4 and 5 will use the information and expand upon the examples presented.

2.2 MATHEMATICAL MODEL OF BICYCLE PERFORMANCE: POWER REQUIRED

The power consumed in propelling a bicycle and rider is primarily due to overcoming wind resistance and lifting mass up hills at normal bicycle speeds and secondarily to bearing and tire friction. Bearing and tire friction, although small, can equal wind resistance at very low speeds.

2.2.1 Hill Climbing Power

The power consumed in climbing a hill (P_u) in watts (W) is the product of total mass and vertical speed:

$$P_u = 9.81 M v_g G \quad \text{(W)} \qquad (2.1)$$

Electric Bicycles: A Guide to Design and Use, by William C. Morchin and Henry Oman
Copyright © 2006 The Institute of Electrical and Electronics Engineers, Inc.

where M is the total mass of the bicycle, its rider, and cargo being carried, in kilograms; v_g is the ground speed in meters per second; and G is the grade expressed as a fraction that is the elevation change over distance traveled. For steep grades G should be replaced by sine $[\tan^{-1}(\text{rise/run})]$.

2.2.2   Power to Overcome Wind Resistance

The most important variable in the power consumption of a battery-powered electric bicycle is wind drag. The force of this wind drag (R_w) in newtons, from [1] is

$$R_w = C_d \rho A v_r^2 / 2 \tag{2.2}$$

where

$A =$ frontal area, m^2

$C_d =$ coefficient of drag. Examples are 0.1 for a streamlined body 0.3 for a passenger car, 0.77 for a recumbent bicyclist, and 1 for an upright bicyclist. More information on this coefficient is provided later in this chapter.

$\rho =$ density of air, kg/m^3

$$\rho = 1.2 e^{-0.143h}$$

where

$h =$ elevation above sea level, km

$e = 2.7183$

$v_r =$ relative speed in air, which is the ground speed (v_g) plus the wind vector (v_w), m/s

That is

$$v_r = v_w + v_g$$

The power consumed (P_w) in overcoming wind drag in each increment of travel is

$$P_w = R_w v_g, \quad \text{(W)} \tag{2.3}$$

$$P_w = [C_d \ \rho \ A (v_w + v_g)^2] v_g / 2 \quad \text{(W)} \tag{2.4}$$

Note that propulsion force required for overcoming wind resistance varies as the square of the relative speed. The propulsion power delivered, where the tire contacts the ground, varies directly with the speed of travel over the ground.

For example, a bicycle traveling at 10 m/s (36 km/h, 22 mph) into a 2-m/s (7-km/h, 4.5-mph) head wind could require 432 W of propulsion power. The same bicycle traveling at 7 m/s (25 km/h, 15 mph) into a 5-m/s (18-km/h, 11-mph) head wind would need 302 W. Both cyclists would feel the same 12-m/s (43-km/h, 20-mph) wind in their faces as they cycle. Incidentally, for each condition the same energy is used. For the same length of travel, the higher power condition arrives sooner than the lower power condition.

Coefficient of Drag of Shapes of Interest   You might use an aerodynamic shape on the front of your bicycle or you might choose to enclose the bicycle. Table 2.1 shows the drag coefficient values for various shapes. Also shown is the possible benefit in terms of the power required to overcome the air resistance of these shapes for a frontal area of 0.4 and 0.5 m^2 in a 20-mph (32-km/h) wind.

Projected Area of the Bicycle Rider   We have on several occasions measured the components of the frontal area of a street-dressed bicycle rider. We did this by enlarging the picture of rider on a bicycle and then measuring the area of the components of the enlargement. In Table 2.2 the columns show the results for a

TABLE 2.1   Coefficient of Drag for Various Shapes

Shape	Drag Coefficient C_d	Power Required (W)[a]	
		0.5 sq m	0.4 m^2
Circular disc[b]	1.12	189	151
Flat plate $L/D = 1{:}1$[c]	1.16	195	156
$\quad L/D = 5{:}1$[b]	1.2	202	162
$\quad L/D = 20{:}1$[b]	1.5	252	202
Bicycle rider[a]	1	168	135
Cylindrical surface[b] perpendicular to flow $L/D = 5 : 1$	0.74	125	100
Hemisphere[b] hollow downstream	0.34	57	46
Teardrop[c]	0.1	17	13
Ellipsoid 1:3[b] major axis parallel	0.06	10	8
Airship hull[b]	0.042	7	6

[a]For 20 mph (32 km/h) airflow. Note: L/D is shape length-to-width or diameter ratio.
[b]From Ovid W. Eshbach, *Handbook of Engineering Fundamentals*, Wiley, New York, 1952.
[c]Reference [3].

TABLE 2.2   Frontal Area of Certain Parts of the Human Body[a] Riding on 26-inch Mountain Bicycle

	Area (m^2)	(%) of total
Head	0.036	7.2
Torso	0.127	25.1
Arms	0.09	18
Upper legs	0.094	18.7
Lower legs	0.053	10.4
Feet	0.025	4.9
Hands	0.026	5.2
Bicycle	0.053	10.5
Total	0.504	

[a]175-lb (80-kg) street-clothed man.

175-lb man in a regular short-sleeved shirt and normal dress pants [2]. The man's image in the enlarged video picture had a measured total frontal area of 0.5 m².

The data shows areas of the body segments so that you can explore the benefit of aerodynamic improvements from placing selected parts of the bicycle rider's body behind aerodynamic shields.

2.2.3   Power to Overcome Rolling Resistance

In traveling at a very low speed on a level route, the force that a bicycle's propulsion must overcome is mainly the rolling resistance. This resistance depends on the vehicle weight, the type of bearings used, and the type of tires. Rather than attempting to determine the many components in this rolling resistance, we measured the total effect [3]. Rolling resistance was measured for travel over a smooth concrete factory floor, where there was no wind. Vehicle weight and tire pressure was varied. We found that the coefficient of rolling resistance varies with vehicle mass.

We approximate the coefficient of rolling resistance, C_r, by the function:

$$C_r = A + B/W \tag{2.5}$$

where W was our measurement of weight in pounds and values of the parameters A and B are given in Table 2.3.

Plotted in Figure 2.1 are the coefficients of rolling resistances of bicycle types that are of various masses and tire pressures. The smooth lines on the graph in Figure 2.1 show the results of these approximations for a three-wheel recumbent vehicle that we used in a classic Seattle to Portland bicycle run. This vehicle had one front wheel with a 2- by 16-inch knobbed tire and two rear wheels with 1.75- by 16-inch smooth-surface tires [3].

We also measured rolling coefficients outdoors in very low wind conditions by pulling the recumbent and measuring the required towing force. We made 30 measurements in both directions on level gravel and asphalt roads at a speed of 2 mph (1.2 km/h) or less. The recorded results are shown in Table 2.4.

The power consumed in overcoming bearing and friction (P_r) effects is

$$P_r = 9.81 C_r M v_g \quad \text{(W)} \tag{2.6}$$

TABLE 2.3   Parameter Values for Eq. (2.5) Coefficient of Rolling Resistance

	A	B
Electric bicycle		
35 psi	0.0031	0.75
Three-wheel recumbent		
15 psi	0.0077	0.53
35 psi	0.002	0.455
50 psi	0.0019	0.423

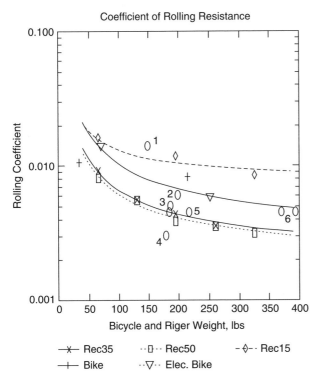

Figure 2.1   Measured values of coefficient of rolling resistance. Combined values are from [2] and [4]. Data points are: ×, three-wheel recumbent with one front wheel having a 2- by 16-inch knob tire, and on the rear wheels 1 $\frac{3}{4}$- by 16-inch smooth-groove tires. All three wheels have 35 psi tire pressure. □, same recumbent with 50-psi tire pressure; Δ, same recumbent with 15-psi tire pressure; +, mountain bicycle with 2 $\frac{1}{4}$- by 26-inch knob tires at 45 psi; ∇, is an electric mountain bike with 26-inch tires at 35 psi; 0, various bicycles as numbered on graph from Gross et al. [4]: **1** is for youth off-road racer, **2** is for European commuter, **3** is for recumbent, **4** is for racing, **5** is for three-wheel aerodynamic, and **6** is for tandem bicycle.

TABLE 2.4   Measured Values of Coefficient of Rolling Resistance for Various Road Surfaces

Weight			
lb	218	282	381
kg	99	128	173
	Coefficient of Rolling Resistance		
Compacted gravel	0.01	0.0052	Not measured
Loose pea gravel	0.0092	0.0069	0.0051
Smooth asphalt	0.0071	Not measured	Not measured

Source: From Morchin [3].

For example, 43.4 W would be needed to propel a 114-kg (250-lb) bicycle with rider at a speed of 5.55 m/s (20 km/h, 12.4 mph), when the tailwind velocity equals the bicycle's forward velocity and $C_r = 0.007$.

2.2.4 Power for Acceleration

The kinetic energy, K_e, invested into a moving object corresponds to one-half its mass times its velocity squared ($K_e = \frac{1}{2}Mv^2$). A propelling force that exceeds the object's windage and friction forces will raise the object's kinetic energy content by raising its velocity. The accelerating force times the distance over which it is applied corresponds to the increase in kinetic energy of the mass.

The force of acceleration is $F = Ma$. This force translates to torque required at the bicycle drive wheel that ultimately requires a torque from the motor and/or a force exerted by the rider on the crank pedals. The torque required at the drive wheel for a given acceleration (a, m/s^2) is

$$T = 9.807Mra \quad \text{(N-m)} \tag{2.7}$$

where r is the wheel radius, expressed in meters, and the bicycle and rider mass that is the combined weight in kilograms divided by the gravitational constant 9.807 m/s^2. For example, a bicycle with 66-cm wheels with a combined weight with rider of 130 kg that is accelerated at 1 m/s will require a wheel torque of $9.807 \times 130 \times 1 \times 0.33/9.807 = 43$ N-m. The gearing between the motor and/or the crank pedals will scale this torque value by the gear ratio. Gears that reduce the speed of the crank pedals and motor relative to the drive wheel will reduce correspondingly the torque required at the motor and/or crankshaft. It is natural when riding a bicycle to start with a push off and pedal force.

The moment of inertia of the bicycle drive wheel, the gears, and the motor rotor and the inertia of other components in the drive system will add to the torque required. However, in comparison, these can be assumed negligible in a practical sense.

The force exerted by the rider on the crank pedals will subtract from that required of the motor. As already stated, it is natural when riding a bicycle to start with a push off and a pedal force. However, if the rider does not contribute to the acceleration force, the motor will bear the full load. In this case at the instant of the start of acceleration, because there is no motion, the motor delivers no power but must supply the torque to start the bicycle to move. The starting torque, depending upon conditions, is likely to be near the stall torque conditions of the motor, at which point the current will be high. The battery must supply this relatively large current at its terminal voltage to achieve the acceleration. This is a condition that requires maximum power from the battery, to the extent necessary for acceleration.

To find the power necessary for this acceleration we can start by determining the work done. Work is the product of force and distance. The distance $S = \frac{1}{2}at^2 + v_0t$, where v_0 is the initial velocity and t is the time of acceleration.

Assume $v_0 = 0$. Using the above example values, $a = 1$ m/s^2 and $t = 1$ s, we find that $S = \frac{1}{2}$ m. The work done is

$$W = FS = Ma(\tfrac{1}{2}at^2) \quad \text{(kg-m)} \qquad (2.8)$$

Power is the rate of doing work, such that $P = W/t$, so

$$P = \tfrac{1}{2} \, 9.807 Ma^2 t \quad \text{(W)} \qquad (2.9)$$

Continuing with the example, the average power at the first second of acceleration of the 130-kg bicycle and rider is 65 W. At 5 s it would be 325 W, at which point the bicycle would have traveled 12.5 m (20 feet).

At the first instance without motion, there is no work, but power is being drawn from the battery. Continuing with the example of 43 N-m of torque, consider a motor characteristic that might have a rating 81 N-m of stall torque at a current of 60 A at 36 V. Using linear scaling we would estimate the current at 43 N-m of torque to be 60 A × 43 N-m/81 N-m = 32 A. This requires 1152 W from the battery. These current and power values are an upper bound on the actual values that will occur due to the internal resistances of the battery, motor, and connecting circuitry and the inductance of the motor. The inductance will limit the instantaneous current, and the circuit resistances will reduce the battery voltage from the nominal rating. With rider assistance the starting current and battery power required will be further reduced or even eliminated with a conservative rider.

The speed of the bicycle with acceleration is $V = a \times t$ and for the example 1 m/s^2 × 5 s = 5 m/s. Total power required will be the sum of power to overcome wind resistance, rolling resistance, any gravity forces, friction, and also accelerating power.

The energy required, using the relationship $K_e = \frac{1}{2}Mv^2$, to accelerate the example 300-lb (130-kg) mass of a bicycle and rider from various starting speeds to higher speeds that are 5 or 10 mph faster is shown in Table 2.5.

Note that more energy is needed to change speed by a given amount when you are traveling faster. This is because the average speed is greater. More power

TABLE 2.5   Energy Used for Accelerating a Bicycle from Various Speeds

Start Speed		Energy Used (Wh) at End Speed	
mph	km/h	+5 mph (+8 km/h)	+10 mph (+16 km/h)
0	0	0.1	0.28
3	4.8	0.21	0.60
6	9.7	0.33	0.83
9	14.5	0.44	1.07
12	19.3	0.55	1.28

Conditions
  Weight 300 lb (131 kg)
  Acceleration 3 ft/s^2 or 2.05 mph/s or 0.9 m/s^2

is consumed because power is proportional to force times velocity. You will need this information later when determining the capacity of the battery needed for your bicycle. This requires estimating how many times you change speed and by how much.

We point out in Chapter 4 that energy can be recovered during deceleration. By reversing the above concept and considering speed decreases in Table 2.5, the tabulated energy values become recoverable energy. This recoverable energy is decreased by the inefficiency of recovery.

2.2.5   Measured Values

We have tested many configurations of battery-powered electric bicycles. We rode one on a basically level bike trail until the batteries were exhausted. This trail, the "interurban trail" connects the cities of Auburn and Kent in Washington State. We rode the bicycle first north, and then south, to cancel the effect of slight slopes and any light wind that existed. We recorded speed, distance, motor voltage, and motor current as we traveled. These data confirmed the analytic expressions that we used in a computer to produce the bicycle travel performance charts in this chapter [2, 3].

2.3   ESTIMATING REQUIRED MOTOR POWER

We use the analytic expressions given in Section 2.2 to illustrate a range of values for the pertinent parameters such as speed, weight, road grade, wind effects, and propulsion power. These can be used to estimate the required motor power. This estimate along with the information given in Chapter 6 can be used to select a motor. The process is first to find the propulsion power, second to scale it upward to account for losses, and lastly to determine how much of the propulsion power is to be contributed by the rider.

2.3.1   Determining Propulsion Power

The above functions determine the power necessary from the drive wheel to move the bicycle at the desired speed. Example results using these analytic expressions for some representative conditions are shown in Table 2.6 and Figure 2.2. Table 2.6 shows the power required for developing the bicycle wheel torque needed for traveling at the indicated speeds, with various surface-grade and head-wind conditions. These values of power are necessary for overcoming (1) the force of air resistance, (2) the force to overcome resistance of the wheel to roll, and (3) the force needed for overcoming gravity when traveling uphill. We used an air density that one expects at an elevation of 50 f in creating the values for the table. Air density at higher elevations will be less, so less power would be required. Adopting a low-elevation value results in more conservative estimates of the needed power. A coefficient of rolling resistance of 0.007 was used.

TABLE 2.6 Power Required to Develop the Necessary Wheel Torque for the Indicated Travel Conditions [a]

		Head-Wind Speed				
	mph	0	6.2	15.5	25	
	km/h	0	10	25	40	
Bicycle Speed						
mph	km/h		Road Power Required (W)			Road Grade (%)
8	12.9	40	69	143	254	0
		156	185	258	369	3
		271	300	374	484	6
12	19.3	86	145	279	468	0
		259	318	452	641	3
		431	491	625	814	6
16	25.7	161	261	472	755	0
		392	492	702	985	3
		622	722	932	1216	6
20	32.2	277	428	730	1124	0
		565	716	1018	1412	3
		853	1004	1306	1700	6

[a] Conditions: Rider and bicycle frontal area is 0.4 m² (4.3 ft²), weight is 75 kg (165 lb). Coefficient drag is 1: coefficient rolling resistance is 0.007; elevation is 15 m (50 ft).

If the rider is bicycling against a head wind, then the aerodynamic loss grows, and at each speed more power must be delivered to the wheels at the road surface. Table 2.6 illustrates this by showing the power required for head-wind speeds up to 25 mph (40 km/h).

For air resistance effects we assumed a frontal area of 4.3 ft² (0.4 m²). This would correspond to an adult person wearing normal street clothes. We measured the projected frontal area from photographs of people riding street bicycles in a normal relaxed position. They were not racers who are normally crouched down over the handlebars. We used a coefficient of drag of 1 that is based on measurements. Also, for the combined weight of the rider and electric bicycle we used 75 kg (165 lb). Figure 2.2 shows the total power required to be delivered at the road surface, in a calm atmosphere, for bicycle speeds from 5 to 35 km/h and road grades from 0 to 9 percent.

2.3.2 Scaling Road Power to Find Required Motor Power

In determining the propulsion motor's rating the, designer must increase the values in Table 2.6 and Figure 2.2 to include power for mechanical and electrical losses. Additional motor power is also required for traveling over a bumpy road or a road that has water puddles. Motors are rated in terms of power output. This output must include the losses in the gears and chain that couple a high-speed motor to the bicycle wheels.

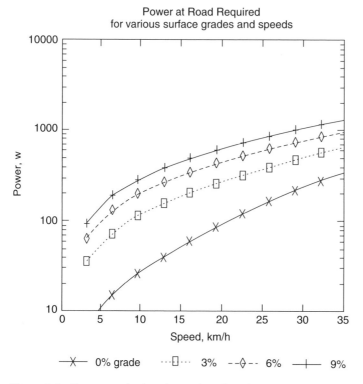

Figure 2.2 Power required at the road surface for various road grades and travel speeds. Legends: x is 0% road grade, [] is 3% road grade, ◊ is 6% road grade, and + is 9% road grade.

A practical way of accounting for these losses is to combine them into an overall efficiency. This would assume, which is normally the case, that the variable losses within the battery and the electrical circuit between the battery and the motor are not as significant as the changes in the losses within the motor whenever its power output changes. Dividing the Table 2.6 values by the overall efficiency, which we refer to as system efficiency, produces a conservative value for the required motor power rating. This power, when divided by 746, gives the value in horsepower. One needs to modify this procedure for determining motor power if the losses between the battery and the motor are not insignificant. This modification would be necessary if a high-loss component, such as a rheostat, is used to control motor speed.

Adequately sized wiring, switches, and solid-state circuits have small losses that become negligible when compared with weight and size differences between bicycle riders.

An additional factor in determining motor power ratings is the input/output ratio of the speed reduction gear that couples the motor to the bicycle power drive wheel. The best speed reduction gear may not be appropriate for the most efficient motor. Selecting the optimum type and revolutions per minute (rpm) ratio for efficient bicycle propulsion is a complex system engineering procedure that is described in Chapter 6.

We have found by experience that total overall propulsion efficiency is around 50 to 60 percent. The motor can be the most significant contributor to the loss of efficiency. Our measured maximum efficiencies range from 30 to 70 percent for brushed series-wound and permanent-magnet direct current (dc) motors. Many motor manufacturers quote efficiencies of 90 percent or more. However, the efficiency of a motor depends on where you operate the motor on its load–power design point. Generally, correct operation of the motor results in about 70 percent efficiency.

Coupling the motor to the drive wheel results in additional loss to system efficiency. Some of the mechanisms used to couple the motor to the wheel are gearboxes, belts, chains, or friction drive mechanisms. A gearbox will reduce the efficiency by roughly 5 percent for each factor of 2 speed reduction. A chain or cogged belt has a high efficiency, on the order of 95 percent. A friction drive device on the bicycle wheel has an efficiency that varies with pressure against the wheel and also wet or dry road conditions. In a trial-and-error adjustment of pressure against the wheel, we achieved a balance between positive wheel drive without slip and minimum current draw from the battery. We found that the highest efficiency is around 85 percent for this type of drive.

Motor power values determined by the above procedure assume that the rider adds no power. Nearly all-electric-powered bicycles can be pedaled while being powered by the motor. The motor power can then be reduced, especially for the higher surface grades. Availability of this pedal power can reduce the required motor power for bicycling up short steep hills.

To illustrate how one would choose the rating of a motor that would deliver the bicycle propulsion power required, we use the computations for the sample expected terrain and rider. The process of selection would be an iterative one wherein a sample terrain would be selected and certain assumptions would be made on the upper limits of motor power and how the rider would contribute to the propulsion force. At each iteration step assumptions would be changed until an acceptable result was obtained.

Figure 2.3 shows an example steep hill, the Graham hill, and the corresponding power required from a motor and from a rider. The computed power values assumed a system efficiency of 60 percent, a bicycle and rider with a combined weight of 120 kg (265 lb), a frontal area of 0.42 m^2 (4.5 ft^2), and the speed values shown in the figure. A maximum motor power was set at 375 W. The total trip time for this illustration was about 15 min for which about 10 min was spent in uphill travel. During the uphill segment, the rider had to produce more than 100 W for 6.5 min or more than 200 W for 3 min.

In Chapter 5 we will discuss how to select a motor according to its ratings and speed–torque curve.

2.3.3 Motors Versus Human Muscles for Bicycle Propulsion

The capability of a human to deliver bicycle propulsion power is a function of the time before becoming exhausted, as shown in Figure 2.4. The data generally

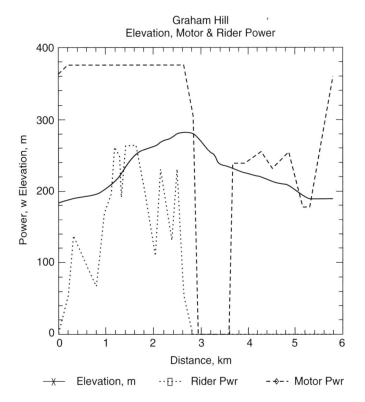

Figure 2.3 Graham steep hill and the power required from the motor and the rider. Graph lines are: _____ elevation in meters, ············ rider power in watts, and - - - - - motor power in watts. Bicycle speed in km/h $= 16 - 1.6\ G^{0.71}$ for $G = 0$ to 6 and $\delta = 24 - 1.6\ G^{1.4}$ for $13 < G < 46$ and 38 km/h for $G < 0$.

pertain to a healthy nonathletic male. We show this only to add perspective and not to be specific with such values. There is a wide range of possible values. The approximate curve shown in Figure 2.4 is given by the equation:

$$P = 85 + 400/\tau^{0.25} \quad \text{(W)} \tag{2.10}$$

where τ is the duration of endurance time in minutes. It is valid only for the extent of the data shown in the figure.

We learned from engineers who are developing the Raven human-powered airplane that a top-notch athlete in good condition can deliver with his legs 4 W/kg of body weight. The Raven's planned flight duration is 5 h. Also, A. C. Gross and his associates report that a healthy nonathlete can deliver 75 W of muscle power for a period of 8 h, while a good athlete can produce about 300 W of power [4]. A typical human can momentarily produce about 746 W. A nonathletic person can deliver this power for 12 s, and the good athlete can deliver it for 30 s.

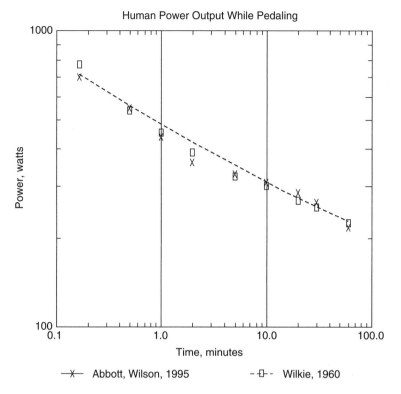

Figure 2.4 Human power output while pedaling. The data points are from: "x" from Abbott and Wilson "Human-Powered Vehicles", Human Kinetics (1995), and the points "[]" are from Wilkie, "Man as a Source of Mechanical Power", Ergonomics, 3 (1), pp. 1–8, 1960, as given by Coate [Ref. 1.1]. The line, - - - - - , in the graph is an approximation.

2.4 SELECTING A BATTERY FOR MINIMUM LIFE-CYCLE COST

We can select the kind of battery to use and calculate how big it should be after calculating what motor power is needed and the length of travel between battery recharges. Battery technology is advancing, and new performance improvements are becoming available every year.

Producers of electric bicycles like to reduce the battery weight by limiting travel range on a fully charged battery. For any given travel range, adopting a lighter but more expensive battery can reduce the battery weight. In Chapter 3 we discuss in detail the characteristics of batteries. Here we offer a method of choosing a battery based on some of these characteristics. The energy density, cost, depth of discharge, and number of charge–discharge cycles available during the battery's useful life are terms we consider for finding the lowest life-cycle cost of battery use.

TABLE 2.7 Battery Life-Cycle Cost Comparisons

	Battery Type					
	Pb	NiCd	NiMH	Li−Ion	Li−Poly	Zn−Air
Energy use per unit distance, E_u, Wh/km	10	10	10	10	10	10
Battery purchase cost per unit of energy, C_e, cents/Wh	4.5	8	35	16	6	0.75
Number of charge−discharge cycles, N	300	1250	1250	1250	1200	200
Depth of discharge (fractional)	0.7	0.65	0.65	0.75	0.7	0.7
Battery cost per unit travel distance, C_m, Cents/km	0.214	0.098	0.431	0.171	0.071	0.054
Number battery purchases relative to longest life battery	4.2	1.0	1.0	1.0	1.0	6.3
Equal travel distance battery cost, cents/km	0.89	0.10	0.43	0.17	0.07	0.33

In Table 2.7 we compare six battery chemistries with respect to cost and available travel distance. The lead−acid (Pb), the nickel−cadmium (NiCd), and the nickel−metal hydride (NiMH) are the most readily available battery types for the builder of an electric bicycle. The zinc−air battery is at this time for illustration only because the battery "charging" infrastructure is not yet in place for bicycle propulsion battery sizes.

Lithium ion (Li ion) batteries that are now being built for electric automobile service might soon be used on bicycles. They can deliver 224 Wh/kg at a cost of $1.22 US per Wh and weigh 4.5 kg/kWh.

2.4.1 Battery Life-Cycle Cost

The total cost of traveling with an electric bicycle is the cost of the equipment plus the cost of electricity used. A bicycle frame and wheels will easily last over 10 years, so their per-year cost is small. Batteries, if deeply discharged every time the bicycle is used, will be the most important factor in life-cycle cost calculations. The effect of depth of discharge on battery life is developed in Chapter 3.

The cost of the batteries and the cost of electricity to recharge the batteries represent most of the cost of travel on a per-unit distance basis. To illustrate, we assume that the electric bicycle uses 10 Wh of electric energy to travel 1 km. Fully discharging a 30-Wh/kg lead−acid battery would carry a traveler a distance (d) per-unit battery weight:

$$d = (30 \text{ Wh/kg})/(10 \text{ Wh/km}) = 3 \text{ km/kg weight of battery}$$

Rather than completely discharge the battery, we might assume a depth of discharge of 80 percent, which delivers 2.4 km/kg of battery weight. A more reasonable limit to depth of discharge for longer battery life would be 60 percent, which delivers 1.8 km/kg of battery weight. More information on depth of discharge is given in Chapter 3.

A general formula that gives battery investment cost (C_m) per unit of travel distance during the battery's lifetime is

$$C_m = E_u C_E / N D_{dis} \qquad (2.11)$$

where

E_u = energy use per-unit distance, Wh/km

C_E = battery purchase cost per Wh

N = number of times battery recharged following discharge

D_{dis} = depth of discharge of battery each time battery is discharged

And D_{dis} is expressed as a factor, with a value less than 1. If N represents the number of charge–recharge cycles of the battery life, one can obtain the relative cost of travel by normalizing C_m to the longest life battery. This is accomplished by multiplying by the ratio N_{max}/N_1 where N_{max} is the battery with the longest cycle life and N_1 is the battery of interest. This ratio determines the number of times the shorter life battery needs to be purchased to provide the same travel distance as the longer life battery. This evaluation is summarized for typical conditions in Table 2.7 for six battery types.

Other factors will enter into the battery cost. For example, a lead–acid battery can be ruined if it is left completely discharged for a few days. A nickel–cadmium battery will withstand more of this kind of abuse.

The total life-cycle cost also includes the cost of the electric energy used in travel. For example, electricity that costs 10 cents/kWh corresponds to 0.01 cents/Wh. With a battery charge efficiency of 50 percent the electric energy cost (C_e) is

$$C_e = 0.02 \text{ cents/Wh} \times 12 \text{ Wh/km} = 0.24 \text{ cents/km}$$

This value or a similar one for your area, when added to the battery investment cost, gives the total life-cycle cost of the battery.

2.4.2   Battery Weight and Volume

Battery weight and volume are important too. It just depends upon how much you want to pay for these other beneficial qualities relative to a more bulky or heavy battery. Formulas for volume and weight are, respectively:

$$V_b = D E_u / D_{dis} E_d \quad \text{(liters)} \qquad (2.12)$$

$$W_b = D E_u / D_{dis} E_b \quad \text{(kg)} \qquad (2.13)$$

where

D = distance

E_b = energy content of the battery, Wh/kg (also referred to as the gravimetric energy density)

D_{dis} = depth of discharge as above

E_d = volumetric energy density, Wh/L

E_u = energy use per unit distance, Wh/km

In Chapter 3 we will show that the energy density depends upon how much power is being drawn from the battery. Less energy can be drawn from a battery, or almost every energy source, when more power is drawn from that battery or other power source.

2.5 UNIQUE NEW TWO-WHEELED VEHICLES

We illustrate the use of some of the analytic techniques developed in this chapter by applying them to two unique and interesting two-wheeled vehicles. The illustrations show the possibilities for evaluating products and designs.

2.5.1 Cheetah — A Superfast Bicycle

From time to time you will come across reports of superfast bicycles. Such superfast bicycles are marvelous works of engineering and craftsmanship. You probably wonder, now that you have seen some of the basic equations of the power it takes to travel, "How did they do it?" The Cheetah, so named by the graduate students at the University of California at Berkeley, is a bicycle that caught our interest. It set a world speed record on September 22, 1992 [5].

This one-page reference did not give much information about how this record-breaking performance was achieved. It showed an aerodynamically enclosed recumbent human-powered bicycle and reported that it traveled at an average speed of 68.73 mph (110.61 kmph) over a distance of 656.2 ft (200 m) and weighed 29.5 lb (13.37 kg). This reported performance had been demonstrated on a roadway in the San Luis Valley of Colorado. The aerodynamic enclosure was reported to be 9 ft long by 18 inches wide. From this information we wanted to know: (1) how much power did it take and (2) what was the aerodynamic drag coefficient?

By scaling the pictures in the reference for the height of the enclosure and of the cyclist, referring to a map, and assuming a moderately built muscular man was the cyclist, we used the following values:

1. The frontal area of the enclosure is approximately ellipsoidal in shape with a minor axis of 18 inches and a major axis of 51 inches, which results in an area of 5 ft^2 (0.46 m^2).

2. The cyclist was about 64 inches tall and weighed about 140 lb.

3. The elevation of the route was about 8000 to 9000 ft.

From the above information and assumptions we calculated, for a level road, that the coefficient of drag was between 0.07 and 0.09. The aerodynamic drag coefficient Table 2.1 shows that this would fall between the teardrop and airship-hull categories. We figured that the cyclist had to deliver 490 to 630 W of power, excluding rolling resistance. An additional 60 to 120 W would have been necessary to overcome rolling resistance defined by an assumed rolling coefficient of 0.0025 to 0.005. This power, which was required at the 68.73-mph speed, was the average during 6.5 s of travel over the measured distance.

You may wonder why Colorado was chosen for the demonstration. The reasons were the high elevation and dry air. The higher elevations have a less dense atmosphere, and hence less air resistance. An additional 270 W of power would have been necessary at sea level to achieve the same speed.

How did the cyclist get up to speed? We don't know exactly, but if the cyclist accelerated at a steady 1 ft/s/s (0.68 mph/s, 0.31 m/s/s), the speed would have been achieved in about 100 s. The power required of the cyclist would have linearly increased from 0 at the beginning to that required at the demonstrated speed. The cumulative energy required was about 10 Wh.

2.5.2   Segway Scooter

We learned about a very impressive test result of the Segway scooter described in Chapter 1. It traveled all the way up an automobile road to the top of Mount Washington in the state of New Hampshire. This road starts at an elevation of 1585 f and rises to 6181 f in a distance of 7.4 miles. Its overall average grade is 11.7 percent.

Being curious about the technical aspect of the Segway, we analyzed the capability of its power train. We made assumptions based on our experience with motors, batteries, tires, and aerodynamics of humans to calculate the energy consumption and power levels required for such a climb. We used our test-verified computer models that previously made similar computations for our electric-powered bicycle designs.

We entered into the computer model the publicized physical aspects of the roadway, including the elevation and distance for each of the 58 segment lengths. The choice of the length of each segment varied and was selected on the basis of departure from a uniform rise in elevation. The average length of a segment was about $\frac{1}{8}$ mile. Thirty-five percent of the segments had a grade slope of more that 19 percent, 46 percent had a grade between 10 and 19 percent, and 4 percent of the segments had a grade between 0 and 5 percent. The remaining segment lengths had downhill slopes.

Other parameters for the modeling were assumed values for rolling resistance of the wheels, weight of the rider, frontal area of the rider, coefficient of drag, standard atmospheric conditions with no wind, system efficiency that varied with road grade, and Segway speed that varied with road grade. The speed of the Segway was assumed to be linear with grade varying from 12 mph (19.3 kmph) at 0 percent grade to 2.5 mph (4.0 kmph) at 25 percent grade. As the motor is loaded, its speed is reduced. The remaining assumptions made are:

Rider weight	165 lb
Drag coefficient	1
Frontal area	7.45 f^2
Rolling coefficient	0.005
Segway stabilization power	0 W

System efficiency ranged from 65 percent at a road grade of 0 percent, to 10 percent at a road grade of 20 percent, and was a constant 10 percent for grades greater than 20 percent. In this demonstration the road grade was such that the combined 4-hp rating of the two Segway motors was exceeded by more than 30 percent in 33 percent of the length segments analyzed. On some occasions the rating was exceeded by 50 percent. The Segway must have had a good thermal design, or we may have been mistaken in our assumptions and in particular overestimated the speeds traveled.

Results of the analysis, summarized in Figure 2.5, show the cumulative energy used for the distance traveled, as well as the road grade and the motor power. The infrequent occurrence of negative grades was used in the computations but not plotted.

Assuming that nickel metal hydride batteries are used and that the battery pack was said to be replaced six times, we find that the total 2490 Wh energy consumption would have been supplied with seven freshly charged sets of batteries. This leads to an approximate battery capacity of 355 Wh. Using typical energy density characteristics for such batteries as 130 Wh/L and 60 Wh/kg, we find that the battery pack weighed about 6 kg, or 13 lb, and occupied a volume of about 2.7 liters, or 167 inches^3. This assumes that each battery pack was fully discharged and that negligible energy was used to stabilize the Segway.

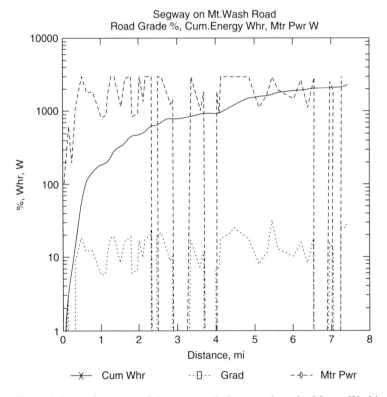

Figure 2.5   Performance of the scooter during travel up the Mount Washington Road.

Travel on level terrain using the same assumptions leads to an energy consumption of 19.3 Wh/mile at 12 mph and 12.1 Wh/mile at 8 mph. Using the 11-mile maximum range, we find that the battery pack supplies 212 Wh for the higher travel speed. Evidently, this is a 60 percent depth of discharge. At the lower speed the Segway should be able to travel 17 miles (27 km).

REFERENCES

1. J. J. Taborek, *Mechanics of Vehicles - Part 6: Resistance Forces*, pp. 26–29.
2. W. C. Morchin, Battery-Powered Electric Bicycles, IEEE Technical Applications Conference, Northcon 94, Seattle Washington, October 12–14, 1994, *Conference Record*, pp. 269–274.
3. W. C. Morchin, Trip Modeling for Electric-Powered Bicycles, IEEE Technical Applications Conference, Northcon 95, Seattle Washington, November 4–6, 1996, *Conference Record*, pp. 373–377.
4. A. C. Gross, C. R. Kyle, and D. J. Malewicki, The Aerodynamics of Human-Powered Land Vehicles, *Scientific American*, September 1983, pp. 142–152.
5. K. Leutwyler, Speed versus Need, *Scientific American*, October 1997, p. 98.

CHAPTER 3

SOURCES OF ELECTRIC POWER FOR BICYCLES

The on-board battery on an electric bicycle is the key propulsion component of this vehicle when reliable performance, long vehicle lifetime, and low per-mile cost are required. Batteries have been used to store energy for decades. Recently, energy-storing batteries for spacecraft, aircraft, and military applications had to meet order-of-magnitude improvements in energy storage capability and lifetime. Batteries using new electrochemical processes were developed and tested in severe environments to prove their life in service. These research and development programs produced significantly improved batteries that are available for propelling electric bicycles. Some of these new batteries are based on complex electrochemical processes and require precisely controlled charging voltages. In this chapter we describe these new developments so that the reader can (1) select the battery that best meets his or her requirements of performance, reliability, and life-cycle cost, (2) specify or design the battery charge control, and (3) identify and correct the causes of deficiencies or failures that could be encountered during the operating life of the battery.

3.1 INTRODUCTION

In the introductory paragraphs we first review the requirements of bicycle propulsion and the performance of batteries that were used in the past to propel bicycles. Then we summarize the topics that must be considered in an analysis in which the new high-performance batteries are compared with the previously used lead–acid and nickel–cadmium batteries. For example, a Hitachi Maxeli lithium ion battery can last for 25,000 charge–discharge cycles. Also its weight is less than one-fourth the weight of a lead–acid battery that has the same energy-storing capacity. On a bicycle that is used occasionally for short-distance travel, the lower cost lead–acid battery might be the best choice. For daily commuting the lead–acid battery might have to be replaced every year. A lithium ion battery could last for decades.

We introduce the subject of battery electrochemistry by summarizing pertinent energy-storing features of the lead–acid battery, which is still being used to start automobile engines. Lead–acid cells powered the bicycles that we had instrumented for measuring propulsion energy consumption in climbing hills that had various grades and in traveling over highways with various road surfaces. However, lead–acid batteries are heavy and their lifetime is limited when they are deeply discharged during each use. Consequently, lead–acid batteries were not suitable for storing energy in space vehicles, so nickel–cadmium, nickel–metal hydride, and nickel–hydrogen batteries were developed for earth-orbiting spacecraft. These spacecraft batteries have been too costly for electric bicycle propulsion.

The need for light weight and long life in batteries that store power in communication equipment has motivated worldwide development programs that are now producing new types of batteries. An example is the lithium–ion battery, which is now in mass production and fills this need. However, lithium cells are very sensitive to temperature, and a cell could be ruined when overcharged. Sophisticated integrated-circuit charge controllers, described in Chapter 4, can prevent battery damage from unanticipated cell temperatures.

In the sections that follow we describe pertinent features of batteries that, when properly controlled, have delivered years of service on successful earth-orbiting satellites. Then we evaluate new batteries that, although complicated, deliver amazing performance that is pertinent to their use in propelling electric bicycles for long travel distances between battery chargings. The best candidate is the lithium ion battery. Among new energy storage systems that have potential for bicycle propulsion are nickel–hydrogen batteries, silver–zinc batteries, and sodium–sulfur batteries.

In the Section 3.3 we describe fuel cell performance. Prototype fuel cell power plants that were fueled with hydrogen were once installed in New York City and Tokyo because their efficiency was higher than the efficiency available in steam-turbine power plants that burned coal. Today's best combined-cycle power plants convert the energy in natural gas to electric power with an efficiency of over 65 percent. Consequently, the development of fuel cells for use in electric power plant use has ended. Fuel cell development was then directed to spacecraft applications, and manned lunar missions were powered with hydrogen-consuming fuel cells that also produced water as a by-product.

The infrastructure that would support a fleet of hydrogen-fueled vehicles would be complicated. A recent development is to deliver ammonia fuel to vehicles. A simple vehicle-carried catalytic unit would release the hydrogen from ammonia. Hydrogen fuel would be inconvenient to carry on bicycles, but the new zinc–air fuel cell, which is being tested in automobiles and buses, might power electric bicycles. Refueling requires simply draining out the zincate-loaded electrolyte and pouring fresh electrolyte and zinc powder into the cell.

3.1.1 Electric Bicycle Battery of 1895

On January 27, 1894, Hosea W. Libbey of Boston, Massachusetts, filed a patent [1] for a zinc–carbon battery that he used in the electric bicycle that he invented in 1897. The cylindrical battery is shown in Figure 3.1. He also had plans for mounting

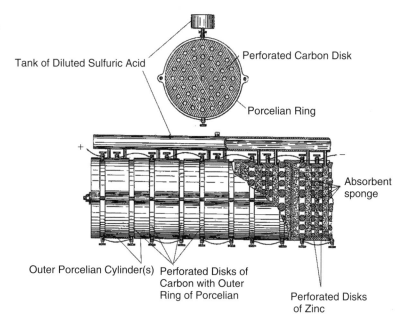

Figure 3.1   Carbon–zinc battery invention from the year 1895 by Hosea Libbey. (After Libbey [1].)

the battery on top of a trolley car, extending from one end to the other. However, in his bicycle invention he shows the battery to have a narrow rectangular-shape bottom and a domed top. It was to fit into the triangular frame of the bicycle between the bicyclist and the front support tube.

Each cell in Libbey's battery contained a pair of perforated carbon disks that were connected to the positive terminal. Each disk was separated, by absorbent sponges, from a pair of perforated zinc disks that formed the battery's negative terminal. The sponges were soaked with dilute sulfuric acid. The acid was stored in a narrow cylinder on top of the battery and was used by the operator as necessary to feed the sponges drop by drop. The assembly would not leak even if the bicycle tipped over.

Periodically, the battery could be disassembled for cleaning of the disks to restore its initial power. Longitudinal bolts passing through tabs on the outside of porcelain cylindrical enclosures held the cells together.

3.2   REQUIREMENTS OF BATTERIES FOR POWERING ELECTRIC BICYCLES

Selecting a battery for a given application is a systems engineering procedure in which all the requirements are defined and quantified. Then all available data pertinent to the alternatives for fulfilling the needs of the application are analyzed. Confirming tests are conducted when necessary. After that the best of the alternatives can be recommended for adoption. This procedure is described in Chapter 6.

The required performance of an electric bicycle affects the battery in which energy is stored. For example, an electric bicycle that is designed for making short shopping trips to grocery stores and carrying heavy loads back to the owner's residence can perform its function with a lead–acid battery.

On the other hand, a lead–acid battery that contains the energy required for day-long travel between charges is not practical on a bicycle because it will be too heavy. A battery used for high-speed everyday commuting 20 miles between home and workplace would also need to carry substantial energy. For this use the per-day cost of a Hitachi Maxeli lithium ion battery could be trivial. In a life test this battery's performance had degraded only by 20 percent after 25,000 charge–discharge cycles. If it supported each day's travel after being charged every evening, then it would have to be replaced on the electric bicycle after 60 years of service.

3.2.1 Requirements

The following requirements need to be quantified for the battery selection process:

- Required electric bicycle performance in terms of travel distance on a full battery charge, allowable battery weight, and power required for traveling when no pedal power is furnished
- Maximum mass of bicycle, rider, and battery
- Steepest hill that must be climbed, its grade and length, the required climbing speed, and how much pedal power could be available
- Range of temperatures in the bicycle's operating environment
- Required minimum battery life in charge–discharge service
- Types of other services for the bicycle might be used, such as local shopping, commuting to work, weekly tours, and long-distance monthly or yearly tours

Other considerations that may affect battery selection include personal desire for ease of travel and importance of maximizing energy cost savings.

3.2.2 Battery Selection Uncertainties

Some data required for an electric bicycle's battery selection process might not be available. For example, Mitsubishi reported a 35,000 charge–discharge cycle life for its 100-Ah lithium ion cells when discharged to 30 percent of its capacity. At the end of life they still delivered 75 Ah in each discharge. However, sophisticated cell balance and cell voltage limit controls were probably used in the test that produced these results.

The integrated circuit battery-charging controls that assure this battery life may be available for the first commercially available cells. Five years later a lower-cost charge control may be adopted. Suppose that the original control fails after 20,000 charge–discharges cycles. A replacement might not be available, so a new battery and controller might have to be purchased. Also the requirements of the bicycle path between home and workplace may change as the gasoline prices rise and more roadways are converted into bicycle paths. The owner might change jobs and move to a different community where he or she bicycles between home and workplace. During its service life the adopted battery might show unexpected loss of capacity that did not occur during its accelerated life test.

The system engineering approach to the possibility of unexpected changes is to specify a battery capacity that is at least 20 percent higher than what the analytical procedure indicates. This causes the bicycle to carry extra battery weight, which might be only a few ounces if a lithium battery is adopted. The extra cost of battery-charging power for propelling this extra weight could be only a few pennies per day.

3.2.3   Measured Energy Comparison in Electric Cars

The world's electric car development programs are producing data that is pertinent to bicycle battery selection. The coming fuel shortage, which is rarely publicized, is accelerating research and developments into alternatives to gasoline-powered transportation in electric automobiles.

The Year 2003 *Michelin Challenge Bibendum*, a 3-day competition among the latest electric cars at the Infineon Raceway in Somona, California, provided useful data on the cost of fuel for electric vehicles. Of the 50 electric cars and light trucks competing in this event, the winner was the tZero, an all-electric two-seat sports car that was built by AC Propulsion [2]. Its battery pack consisted of 8600 lithium ion cells, each cell being 18 mm (0.71 inch) in diameter by 655 mm (2.58 inches) long. The tZero won the travel speed test by going 100 miles (180 km) at 50 mph (80 km/h). In another run it had traveled 280 miles (450 km) on one battery charge. It can travel 153 miles (246 km) by using 33.8 kWh, which is the energy in 1 gal of gasoline.

This electric vehicle obviously reduces the cost of travel. However, it still contributes to air pollution and the nation's petroleum consumption because petroleum is a common energy source for electricity generation.

The 21.7 kWh of electric power required for the tZero's economy run could have been generated from fuel oil that releases 146,900 Btu of heat per gallon burned. The energy in 1 kWh corresponds to 3410 Btu, so 1 gal of oil can produce 25.8 kWh from a 60 percent efficient power plant. Thus, in order to deliver the 21.7 kWh that the tZero needed for traveling 101 miles at 45 mph, plus power transmission and distribution losses, the power plant had to burn and release carbon dioxide from at least 1 gal of oil.

We can conclude that electric cars can reduce the quantity of petroleum that is being consumed by automobile travel in a nation. Petroleum consumption for travel would be zero if battery-powered vehicles were refueled with power from nuclear power plants. Replacing automobiles with electric bicycles would reduce petroleum consumption significantly, even if their battery recharge power came from oil-fired power plants. The electric bicycle can travel about 2100 km

(1300 miles) on the 21.7 kWh that the petroleum-burning power plant generates from burning 1 gal of petroleum.

3.3 CHARACTERISTICS OF BATTERIES SUITABLE FOR ELECTRIC BICYCLE PROPULSION

The key component of a useful electric bicycle is the battery in which energy is stored for supplementing pedal power. The available battery types range from lead–acid, which can deliver 20 Wh/kg of battery weight, to new lithium ion batteries that can deliver 160 Wh/kg. Research laboratories have built lithium cells that deliver up to 400 Wh/kg [3]. SAFT America has developed its type VLE cell that features a specific energy of 195 Wh/kg. AGM Batteries Limited has obtained 235 Wh/kg in its 6-Ah 36550 cells that are based on stabilized lithium metal powder (SLMP). In-cell electronic overcharge protection detects and deviates the charge current when the cell becomes fully charged [4].

Many factors must be considered in selecting the battery type for an electric bicycle because the requirements are complex. For example, battery failure during travel through heavy traffic or climbing a steep hill could be a serious event. A battery with a short life in charge–discharge cycling would need frequent replacement, which raises its life-cycle cost. A heavy battery requires energy for hauling it up hills on the bicycle. This energy has to be furnished by the bicyclist's muscles or from energy carried in the battery. Long battery life is achieved in lightweight batteries by a sophisticated charge control circuit that prevents overcharging cells in the battery. A presently available battery might be replaced in the market by an even better battery, which requires a sophisticated charge control. The older charge control might cease to be available.

Complex problems, such as the selection of a battery type for an electric bicycle, are best solved by an aerospace decision-making procedure called "system engineering," which is described in Chapter 6. The first step in this procedure is to quantify all of the requirements of the system that is being developed. The second step is to create alternative designs that fulfill these requirements with the required performance and reliability and quantified risks identified for each design. The third step is calculating the life-cycle cost of each design and quantifying all of the risks involved. The best option, with risks cited, is then offered to the decision-making authority.

3.3.1 Battery Candidates for Electric Bicycle Propulsion

The world's space exploration and satellite communication programs have supported intense development of high-reliability lightweight batteries for aerospace vehicles. For example, the early Earth-orbiting satellites used nickel–cadmium batteries. They had a short life, about 3000 charge–discharge cycles if 50 percent of their energy content was extracted during each discharge. Replacing a useless satellite required a new satellite that cost $30,000 a pound to launch. Years of testing at the U.S. Naval Research Laboratory showed that the

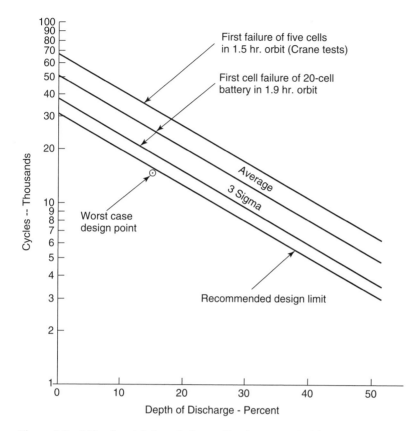

Figure 3.2   Life of a nickel–cadmium cell, when measured in charge–discharge cycles, is reduced when the battery is deeply discharged in each cycle. (From H. Oman, *Energy Systems Engineering Handbook*, Prentice Hall, Englewood Cliffs, NJ, 1986.)

life of the nickel–cadmium batteries could be extended to the life of the satellite by discharging only 15 percent of the nickel–cadmium battery's stored energy during each sun-occulted period of the satellite's orbit (Fig. 3.2).

An early significant product of this spacecraft power research was the nickel–hydrogen battery, which can be fully discharged in each sun-occulted period. A life of 37,000 charge–discharge cycles has been obtained. Nickel metal–hydride batteries followed, and by year 2004 lithium polymer batteries were qualified for space use. Hydrogen/oxygen fuel cells are used in manned spacecraft and in the International Space Station.

A new need for a lightweight high-reliability electric energy source came from manufacturers of hand-carried computers, communications equipment, and portable navigation aids. The availability of Earth–satellite data links produced a big market for these lightweight communication and navigation products. The consequent need for lightweight high-reliability power sources for these applications motivated intense worldwide lithium battery development programs.

Factories are now mass producing small lithium ion cells that have high specific energy content and long life in charge–discharge service.

A popular product of this development is the type 18650 lithium ion cell, which is produced by many factories. It is 18 mm (0.71 inch) in diameter by 655 mm (2.58 inches) long. Its typical energy content, when fully charged, is 169 Wh/kg. These cells are being assembled into batteries for other applications that even include electric cars such as the tZero.

The best battery candidates for storing electric propulsion energy for electric bicycles are listed in Section 3.2.7. Lead–acid batteries propelled electric cars in the beginning of the twentieth century and still are used for electric lift-truck propulsion. They are also used for delivering power that starts automobile engines. However, their higher weight, which corresponds to 20 to 40 Wh/kg, makes them not as practical for most bicycles as other higher energy batteries. Also, their typical guaranteed life for automobile use is typically only 38 to 60 months from date of purchase.

Charge Control of Long-Life Bicycle Batteries Vented lead–acid batteries in the past had acceptable life for charge–discharge service in homes, automobiles, and airplanes. Overcharge would electrolyze into hydrogen and oxygen the water that diluted the sulfuric acid in the battery's electrolyte. These hydrogen and oxygen gases would pass through vent holes in the cell caps into the atmosphere. The battery owner had to periodically check his battery with a hydrometer and add water when the electrolyte was low. Battery life depended on this procedure for ensuring the quality of the battery's electrolyte.

The battery industry responded to this maintenance requirement by developing the *sealed lead–acid battery*. Each cell had a pressure-relief valve in its cap. During overcharge, the electrolysis process produced hydrogen from water at the negative electrode. This hydrogen then drifted to the space above the electrolyte. The hydrogen gas pressure increased until it was high enough to vent the hydrogen through the pressure-relief valve in the cell's cover. The oxygen produced at the positive porous lead plate would drift through the electrolyte to the negative lead electrode. However, the fully charged negative electrode was already covered with lead oxide and it could absorb no more oxygen. This surplus oxygen then drifted back through the electrolyte to the positive electrode, where it entered the pores of that electrode and plugged them with lead oxide. This decreased the sealed battery's energy storage capacity to a level at which the battery had to be replaced.

Lead–acid batteries that are used to store energy in electric utility installations have controls that sense overcharge and stop recharging before overcharging can begin electrolyzing the electrolyte. However, carrying on a bicycle the relatively heavy lead that is in a lead–acid battery is not as practical today as it once was. We can now store electrons in much lighter metals like lithium. Sophisticated integrated circuit components can effectively control the lithium ion battery's charging and discharging.

Evaluating Energy Storage Alternatives for Bicycle Propulsion The most important component in a battery-powered electric bicycle is the battery that supplies propulsion power. A wide range of battery electrochemistries is available for selection in this application. Each electrochemistry has unique features that affect a battery's (1) energy content in watthours per kilogram, (2) service life in terms of number of charge–discharge cycles, (3) first cost and life-cycle cost, and (4) unique features such as ability to survive in a discharged condition.

In the sections that follow we first review (1) the pertinent characteristics of the best batteries for electric bicycle propulsion, (2) the adopted test procedures, and (3) the limits on the applicability of each battery type for electric bicycles. We then search other battery types to find the best batteries for meeting various range and performance requirements of electric bicycles.

An important element in battery-powered electric cars is the overall efficiency that includes the efficiency of the battery charging, as well as the efficiency of its battery for storing energy for the motors that drive the car's wheels. All of the power losses add to the cost that must be paid to the electric utility that supplies the power. If solar power were used to recharge the batteries, the costly area of the solar panels would have to be large enough to deliver all of the systems energy losses as well as the energy that propels the vehicle.

The evaluation of candidate batteries for propelling an electric bicycle involves many factors that are not publicized by bicycle manufacturers. For example, the battery that controls an automobile's starting, ignition, and lighting is probably the lowest cost battery available. However, its sales literature would not reveal its under-100 cycle lifetime if it were completely discharged in every charge–discharge cycle.

The research and development results from well-funded battery development programs are useful tools for the designer of an energy storage system for an electric bicycle. In the paragraphs that follow we summarize the data that is pertinent to electric bicycle design, construction, and performance.

A few general conclusions of early research in this aerospace energy storage program were:

- A 20-year lifetime of new energy storage apparatus is difficult to establish in a 1-year test.
- The starting–lighting–ignition battery used in automobiles has a lifetime, when discharged deeply, of 100 charge–discharge cycles.
- Nickel–iron and pocket-plate nickel–cadmium batteries can be built to last at least 30 years.

Definitions of Battery Types, Components, and Performance Batteries were developed in many countries, using different electrochemistries, and for various purposes. Consequently, battery performance and life are often described in unfamiliar terms. In the paragraphs that follow we describe the terms that are used to define battery performance and what battery components affect performance of different battery types.

Efficiency The *efficiency* of energy storage is the energy recovered from the storage mechanism divided by the energy delivered to it. For example, in a public-utility peak-shaving battery the losses must include the losses in the transformer and rectifier that charge the battery, the inverter that makes alternating current (ac) out of the direct current (dc) battery output, as well as the battery and control losses.

Coulombic efficiency is the ampere-hours delivered by a battery divided by the ampere-hours required to recharge it.

Storage Battery   A *storage battery* is a combination of positive electrodes, which are sometimes called *cathodes* during discharge, and negative electrodes, which are called *anodes* during discharge. These electrodes, plus an *electrolyte*, are arranged to accept charge current from a dc power source and deliver power to a load when discharging. A *battery cell* is a unit that has one or more negative plates in parallel and one or more positive plates in parallel, all immersed in a common electrolyte. Two or more cells connected in series is called a *battery*. A single cell is also often called a battery. Among the smallest storage batteries are those mounted in circuit boards of computers for keeping alive the random-access memories. Some of the largest batteries are installed on submarines to supply propulsion power while the vessel is submerged.

Storage batteries have evolved to match requirements of applications. Batteries designed for one application are not necessarily suitable for another one. For example, a starting–lighting–ignition battery wouldn't last long in traction service. Important energy features of batteries are:

- A battery can have close to 100 percent coulombic efficiency in that every coulomb of charge is recovered during discharge. However, some batteries need to be frequently overcharged.
- The charging voltage is always higher than the discharging voltage at a given state of charge. The voltage difference, which depends on current density, represents a loss.
- The life of a battery, in terms of number of charge–discharge cycles, generally varies with depth of discharge in that deep discharges shorten cycling life.
- Battery price is affected by the cost of materials as well as the technology of manufacturing. The price of lead has varied by a factor of 2 during a year.

Energy Capacity of Batteries   Although one would like to define the *energy capacity* of a source as a given fixed value, it is not. It seems that for every energy source imaginable, the quicker you draw energy from it, the less it can totally supply. This includes humans who can deliver a quantity of energy for a time inversely proportional to the amount of power delivered. This is also true of batteries, fuel cells, and electric generators.

Battery capacity is commonly measured in ampere-hours at a given temperature and discharge rate. Often the current capacity of a battery is quoted as a C/x, where C is the ampere-hour capacity of the battery and x is in hours. If a constant current of 50 A will completely discharge a 500-Ah cell in 10 h, then this current is called a $C/10$ rate. Available capacity over a discharge period, typically for 6, 8, or 20 h, is stated as the ampere-hour rating. Battery capacity is usually affected by temperature, with capacity dropping as temperature is lowered. Also pertinent to a battery's capacity designation is the lowest end-of-discharge voltage that can be accepted.

Sometimes a lead–acid battery is said to reach its end-of-life point when its available ampere-hour capacity is 80 percent of rated capacity. A battery designed for cycling service may improve in capacity during the first few hundred cycles, so at times its actual capacity can be more than its rated capacity.

Depth of discharge is expressed as a percent of the ampere-hour content of the battery that is extracted during one discharge. For example, a 60 percent depth of discharge takes 60 Ah out of a 100-Ah battery, leaving 40 Ah in the battery.

One can expect to draw electric current from a battery at a given ampere rate over the time period that depends on the battery's rating. The battery will deliver fewer ampere-hours if more current is drawn than the rated value. This is known as the Peukert's effect. A simplified version applicable to lead–acid batteries, excluding temperature effects, is

$$C_p = I^n T \qquad (3.1)$$

where C_p is the capacity in ampere-hours, I is the current drain in amperes, n is Peukert's exponent, and T is the time in hours. R. L. Proctor's experience [5] is that a flat-plate lead–acid battery will have an n of 1.25 to 1.3. A gelled battery and some spiral-plate batteries will have an n as low as 1.1. A plot of this equation with the terms normalized to percent of battery capacity is shown in Figure 3.3.

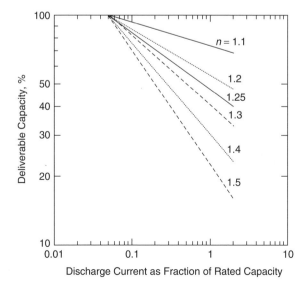

Figure 3.3   Battery capacity versus discharge rate based on Peukert's effect. (After Proctor [5].)

The discharge current that is drawn by the motor on the electric bicycle will vary according to motor load depends upon many variables. Wind, weight, and road grade are the principal ones, as we show in Chapter 2. In general, it would seem undesirable to operate the battery at 10 percent of its ampere-hour capacity. The result would be a long battery life and long travel distance but at the cost of high weight. The practice has been to operate in the region of 50 to 100 percent of capacity for level travel. One will notice that in electric bicycle advertisements the travel distance is oftentimes equal to the value of speed. (This seems to be true for electric cars too.) This implies that the battery is used up in 1 h. Thus, the battery is being used at the approximate rate equal to 100 percent of its capacity. A battery designed for cycling service may improve in capacity during the first few hundred cycles, so at times its actual capacity can be more than its rated capacity.

We see from Figure 3.3 that for gelled lead–acid batteries the deliverable ampere-hours can be about 75 percent of the battery capacity rating. This does not represent the energy deliverable because the voltage is dropping as the current is being drawn. The product of the two integrated over time represents the energy that is deliverable. This is not an exactly predetermined value because of the variable load and temperature conditions during use. One can refer to charts similar to that shown in Figure 3.4. They illustrate the available energy normalized to battery weight for specific power values drawn from the various types of batteries shown.

Using the material in Chapter 2, one can estimate the total energy desired for achieving a particular travel range. With this value one can select the battery type and subsequently the initial trial-battery weight. Knowing the bicycle use will determine the approximate average power to be drawn from the battery. It will probably be in the 100- to 150-W average-value region. Using the initial

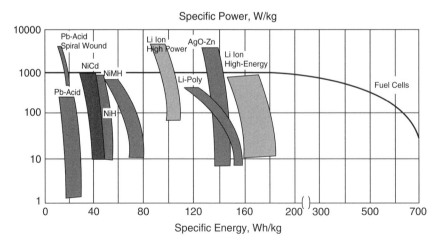

Figure 3.4   Energy storage density vs. power density the Ragone plot. (From Kenneth A. Burke, Fuel Cells for Space Science Applications, in Proceedings of International Energy Conversion Engineering Conference, Portsmouth, VA, AIAA 2003–5938.)

battery weight and dividing it into the average power value determines where to enter the specific power axis of the graph in Figure 3.4. The intersection with the battery curve determines a specific energy. If this value is not equal to that initially determined, one needs to iteratively repeat the process.

3.3.2   Lead–Acid Battery — Its Limits for Electric Bicycle Propulsion

Lead–acid batteries, which propelled the first nineteenth-century automobiles and even bicycles, are the most common and lowest cost energy storage products available today. Virtually all automobiles, buses, and trucks carry lead–acid batteries for engine starting and powering lights when the engine is not running at night. These batteries were used to start automobile engines after Charles F. Kettering invented the electric starter in 1912, and they are still used in most engine-powered vehicles. The huge replacement market for these batteries enables their manufacturers to maintain an efficient production line that keeps this battery's replacement price low. Many longer lifetime engine-starting batteries have been invented, but their manufacturers have not been able to compete with lead–acid batteries for automobile engine-starting service.

The new requirements of energy storage in missiles and spacecraft motivated intense development programs in energy storage. Lead–acid batteries propelled torpedoes that were launched from submarines. Crew members were available to continuously monitor the charge status of the torpedo battery, making failures of torpedo propulsion power rare. Missiles such as the Minuteman and Peacekeeper required high-reliability batteries that were ready for launch at a command that came after many years of storage in a launch silo. The successful batteries for these missiles are not applicable for use in bicycle propulsion because the missile flight time was short, and after-flight recharging of their batteries was not needed.

The person who is designing a new electric bicycle, or is just selecting a battery for electric bicycle propulsion, needs to understand the characteristics of candidate batteries that are pertinent to his application. For example, the best lead–acid battery can store 20 Wh/kg of battery weight for propulsion. Its performance, if derived from related test data, can be confirmed by electrochemical analysis. A supplier of a new type of a lithium battery might claim an energy storage capability of 150 Wh/kg. However, the offered test data might not include the performance of this battery at the variable discharge-power rates that are required in electric bicycle propulsion. Application of electrochemical analysis in evaluating this data could reveal that a given battery has reasonable probability of meeting the electric bicycle propulsion requirement, or that there is very little probability of meeting this requirement. The designer could then either reject this new battery or include appropriate testing in his development program.

One simple example illustrates the range of performances that a bicycle designer must evaluate in his designing process. A typical lead–acid battery that is completely discharged during every bicycle trip can last for 100 charge–discharge cycles. Then it has to be replaced with a new battery. A more costly lithium ion battery could last for 20 years in the same service. Its per-year cost in electric bicycle service could be far less than the per-year cost of a heavier lead–acid battery from which only 30 percent of its energy content is consumed before it is recharged.

Decades of lead–acid battery development have produced an electrochemical analysis procedure that can enable a bicycle propulsion designer to use limited test and theoretical data for predicting the battery performance in electric bicycle service. In explaining this electrochemical analysis, we illustrate its use with its application in predicting the performance of lead–acid batteries. For example, lead–acid batteries have been charged and discharged over a wide range of internal temperatures. The electrochemical analysis can now reproduce accurately the observed battery performance at these temperatures.

To enable the reader to understand potential causes of battery deficiencies, we first review the electrochemical processes that occur in a lead–acid battery as it is charged and discharged. We then describe the causes of failure in lead–acid batteries, and how manufacturers of these batteries for nonautomotive services have been able to develop lead–acid batteries that have reasonably long service life in specific applications such as electric bicycles.

Lead–Acid Batteries: Uses and Performance Limits   Lead–acid batteries are assembled from low-cost materials and have served historic energy storage applications well, but new alternatives are becoming available for replacing lead–acid batteries for many energy storage needs. The following examples illustrate the range of past battery applications.

Lead–acid batteries can be built with low-resistance internal current paths by using thick lead conductors. This construction enables them to deliver high power for the brief period in starting automobile and airplane engines. Their high weight was not a significant problem when carried in wheeled vehicles. They were also used to store energy for critical loads that had to be operated during electric utility power outages.

Lead–acid batteries were also used to propel forklift trucks that transported cargo for short distances in factories and warehouses. The battery weight was useful in counterbalancing the weight of the load being carried on the forks of the vehicle. Small battery-powered carts were also built for carrying personnel in large factory buildings. These batteries could be recharged frequently, so they didn't have to support a long travel range.

A hybrid automobile can deliver good performance in miles per gallon of gasoline consumed because (1) the engine does not have to be sized for an infrequently needed peak power performance, and hence is lighter in weight, and (2) the engine can be run at constant power when needed and can be shut down every time the car stops. Intense and well-financed lead–acid battery development programs for hybrid propulsion were started in the United States. However, hybrid cars developed in other nations used new nonlead high-performance batteries, and they delivered better miles-per-gallon fuel economy. Also, lead–acid batteries still have limited lifetimes in charge–discharge cycling and quickly degrade if left discharged.

This history of lead–acid batteries indicates that they need to be carefully evaluated when they could be used to propel electric bicycles. For a few miles of daily bicycle travel in local service the lead–acid battery, even with its limited life, might be the lowest cost energy storage component. For long-distance travel the lead–acid battery would not qualify as the source of propulsion power because it has to store the energy needed to propel its own mass as well as that of the bicycle and rider. Also, the battery, if only partially discharged, would have to be replaced every 2 or so years because it has lost energy storage capacity. One of the new lithium batteries could last many years in daily charge–discharge service.

In the paragraphs that follow we summarize the pertinent bicycle propulsion characteristics of currently available lead–acid batteries.

Electrochemical Activity in Battery Charging and Discharging   A lead–acid battery is a box that contains series-connected cells. A 2.2-V cell of a lead–acid cell is an assembly of positive and negative plates. Each plate consists of a lead grid onto which are pressed lead particles that create a large porous surface area on which electrochemical reactions take place. Every positive plate is bonded to a lead bus that connects these plates to the positive terminal of the battery. The negative plates are likewise bonded to a lead bus that connects them to the battery's negative terminal. An insulating membrane, through which only hydrogen ions can flow, is placed in every space between a positive and negative

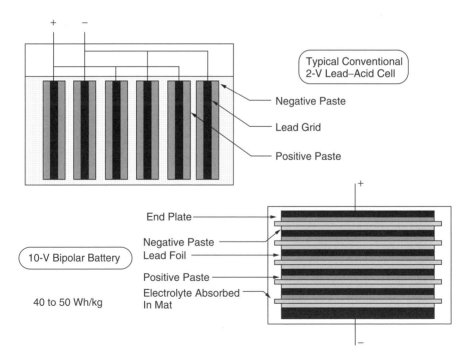

Figure 3.5   Conventional vs. bipolar Lead–acid battery. (From John A. Wertz, Development of Bipolar Lead-acid Batteries with Cladded Materials, IEEE Proceedings of 15th Annual Battery Conference on Applications and Advances, January 11–14, 2000.)

plate (Fig. 3.5). A lead–acid battery is composed of series-connected cells that are in separate cases. For example, a 12-V battery has 6 cells.

After a cell has been assembled, it is filled with dilute sulfuric acid (H_2SO_4), which forms lead–sulfate on each cell plate. Connecting a positive terminal of a 2.4-V charger to the positive post of the cell, and the negative charger terminal to the negative post of the cell, then charges a lead–acid battery cell. This current flow from the charger causes the sulfate on the positive plates to decompose into lead dioxide that coats the plate. This current also increases the density of sulfuric acid in the electrolyte. A positive-charged hydrogen ion is also released. The negatively charged plate attracts this ion, which drifts through the ion-transparent separator to the negative plate of the cell. There it combines with the lead sulfate on the surface of the plate to form sulfuric acid, thus leaving a clean lead surface plate. The negative electron from the ion departs into the circuit that carries it to the negative terminal of the charger.

A battery charger's output voltage is normally limited to a value that converts the entire lead sulfate content on the positive terminal plated to into sulfuric acid. If the battery charger's voltage were not limited, the charging current would then electrolyze the water in the cell into hydrogen and oxygen. This would deplete the available electrolyte, and in sealed cells could raise the internal pressure to damaging levels.

A lead–acid battery is unique in that both plates are at least partly lead, and, when the battery is discharged, both plates are covered with lead sulfate. During the cell discharge, the external circuit receives electrons from the negative plates of the battery. They are produced by the sulfuric acid forming lead sulfate on the surfaces of the negative plates. These plates also release positive hydrogen ions that drift through the separators to the positive plates that are absorbing electrons from the load. These arriving electrons combine sulfuric acid with lead oxide to form lead sulfate on the positive plates. The battery becomes completely discharged when no more lead oxide surface is available for this reaction. A positive plate is normally composed of lead particles that are pressed together to create a surface area that is around 50 times the product of a plate's length and width, times 2 if both sides of the plate are in the electrolyte. Note that during discharge, for every pair of electrons that circulate through the external circuit, one lead atom in the negative electrode and one in the positive plate becomes lead sulfate.

During charge, the negative plate is reduced to pure lead, and the positive plates become lead oxide (Fig. 3.5). Charging current restores the discharged condition of a lead–acid battery by converting the positive-plate surfaces to lead dioxide and the negative plate surfaces to pure lead. After full charge is reached, at 2.5 V per cell in an ordinary lead–acid battery, the cell becomes an electrolyzer, breaking down the water into oxygen at the positive plate and hydrogen at the negative plate. There oxygen combines with the pure lead and sulfuric acid, in effect discharging the negative plate so that it cannot liberate hydrogen. The battery with overdesigned negative plates emits very little gas during overcharge, so it can be sealed, forming a maintenance-free battery.

The sulfuric acid used in lead–acid batteries is diluted to achieve a specific gravity of around 1.250 in a fully charged battery. In some starved-electrolyte batteries the electrolyte has a specific gravity of 1.310 at full charge, dropping to 1.150 at the end of discharge. The state of charge of a lead–acid battery can be inferred easily from the electrolyte density, a feature not available in nickel–cadmium or other cells with alkali electrolyte.

The electrochemical activity in charging and discharging lead–acid batteries is described in greater detail on pages 379–384 in [5].

Lead–Acid Battery Improvements and Limits Long service life is available in lead–acid batteries, but at a price. The key to a lightweight vehicle battery is exposing as much lead surface as possible to the sulfuric acid for the electrochemical reactions. Manufacturers have ways of making a spongy lead plate that has lots of surface area but is so weak that it has to be supported by a lead grid to form a plate. The grid also functions as a current collector. Grid thickness and the alloy used affect the life of the battery. Pure lead resists corrosion but is too weak structurally for most batteries. Alloying the grid lead with calcium or antimony gives the grid strength, but the alloying affects battery performance and life. For example, antimony alloying in the negative plate promotes hydrogen formation.

A lead–acid battery can be designed for low production cost. Its grids would have minimum cross-section area, consistent with allowable voltage drop at the specified output current. Lowest cost materials would be used for separators, and the strongest alloys would be used for lead parts. On the other hand, batteries with pure lead plates lasted for over 30 years in standby telephone service. These batteries are big and costly for their ampere-hour capacity. Long battery life can be obtained at reasonable cost with materials and technology now available. For example, J. S. Enochs was able to get a battery life of 20,000 charge–discharge life with 80 percent depth of discharge in each cycle [6]. He had used nonantimonial lead in his electrodes.

Adapting Lead–Acid Batteries to Electric Vehicle Propulsion Lead–acid batteries had been used for propelling electric cars, but the easily refueled gasoline engine became, and still is, the most popular power source for propelling automobiles and trucks. Also, the useful service life of a lead–acid battery has been limited. For example, lead–acid batteries that start automobile engines generally carry a life guarantee of around 5 years. They generally fail before the guarantee expires, so the user gets credit for a few months of unused life when he or she buys a replacement battery. In contrast, the Edison nickel–iron batteries lasted for many decades of charge–discharge cycling in forklift truck service.

The recent introduction of hybrid cars in foreign markets motivated an intense development of lead–acid batteries for this application. Dozens of papers at each battery conference described innovations and tests that showed improvements in the charge–discharge cycle life of lead–acid batteries and promised even better performance. Then came the development of lithium ion and other new batteries with long life and low-weight performance that no lead–acid battery can match.

Examples of the potential use of lead–acid batteries for short-distance propulsion are the Ford Motor Company's vehicles developed by its THINK Technologies organization. The products were the Think Neighborhood electric car and the Think electric bicycle that was powered by a 24-V gel-cell lead–acid battery. However, these vehicles did not go into production. Neither did the hybrid cars that were developed and tested by General Motors. The lead–acid battery is not the best choice for propelling electric bicycles because of its heavy weight.

3.3.3   Nickel–Cadmium Batteries

A nickel–cadmium cell is commonly made in two forms: (1) a round unit containing rolled electrodes and separator and (2) a prismatic unit in which plates and separators are stacked. The electrolyte is water that contains 25 to 35 percent dissolved potassium hydroxide. The cells can be sealed, but most cells are intended for terrestrial uses so they have pressure-relief devices.

Energy features of nickel–cadmium batteries follow:

- A nickel–cadmium battery will store up to twice the energy that can be stored in a lead–acid battery having the same weight, but the nickel–cadmium battery costs more.

- A pocket-plate nickel–cadmium battery with concentrated electrolyte will have output at $-40°F$, a temperature at which a lead–acid battery is not usable.

- Battery service lives of over 30,000 charge–discharge cycles have been achieved with low depths of discharge.

- The variable charge voltage required for nickel–cadmium batteries makes them inconvenient for service where they float on the line to which they supply power when main power source fails.

A useful source of performance data for nickel cadmium batteries is the NASA Applications Manual, [7]. The cell voltage for various C rates of discharge is shown in Figure 3.6 as a function of nominal ampere-hours delivered.

Nickel–Cadmium Battery Design   The typical nickel–cadmium battery has a positive electrode consisting of a sintered nickel substrate coated with nickel hydroxide. Absorbing electrons from the external circuit during charge converts the surface to NiOOH. Each electron converts one molecule.

The negative electrode also has a nickel substrate but is covered with cadmium metal when charged. Discharging converts the surface to cadmium hydroxide, releasing two electrons from each cadmium atom converted. The chemical equation is:

$$Cd + 2NiOOH + 2H_2O \rightleftharpoons Cd(OH)_2 + 2Ni(OH)_2$$

(Charged)   KOH   (Discharged)

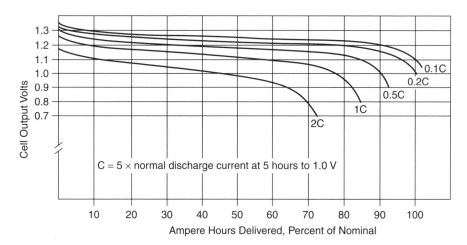

Figure 3.6 Terminal voltage of a nickel–cadmium cell drops as its ampere-hour content is delivered during discharge. (From H. Oman, *Energy Systems Engineering Handbook*, Prentice Hall, Englewood Cliffs, NJ, 1986.)

Energy Research Corporation designed for the U.S. Bureau of Mines a nickel–cadmium battery that replaces the lead–acid battery used since the 1930s for powering miners' cap lamps [8].

Nickel–Cadmium Performance at Low Temperatures Nickel–cadmium batteries can be used at temperatures that are too low for lead–acid batteries. However, the charge and discharge voltages vary with temperature, so a battery operating at variable temperatures requires a charger that senses battery temperature and biases charge voltage appropriately. Also, the loads must accept the voltage that varies with temperature as the battery is discharged. The charge and discharge voltages for a General Electric 1.5-Ah cell are mapped with temperature as a parameter in Figure 3.7.

Charge–Discharge Cycle Life of a Nickel–Cadmium Battery The manufacturer rates a battery on the ampere-hours that it can deliver before reaching a cutoff voltage, often 1.0 V, at a rate calculated to exhaust the battery in a specified time. When delivering its rated content, the battery's storage capacity can be 55 Ah/kg. However, the battery's charge–discharge cycle life would be short if it is discharged fully each time.

Sid Gross, a spacecraft battery specialist, uses curves, such as in Figure 3.2, for selecting battery size. For example, if he needs a battery for a low-altitude spacecraft that orbits the Earth once every 90 min for 3 years, the required service life (L) for the battery would be 17,520 cycles.

From Figure 3.2 he finds that a 14 percent depth of discharge is consistent with the required life. Using a battery rated 31 Wh/kg gives him an actual performance 4.3 Wh/kg. Such is the price for long life. For greater depths of

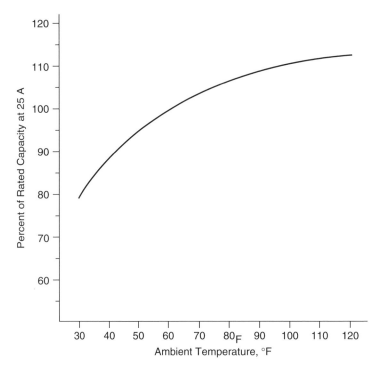

Figure 3.7 This General Electric cell had been designed to operate at temperatures as low as −20°C. (From H. Oman, *Energy Systems Engineering Handbook*, Prentice Hall, Englewood Cliffs, NJ, 1986.)

discharge an approximating relationship for cell life (L) is:

$$L = 10^{(4.7 - D/42)} \tag{3.2}$$

where D is the depth of discharge in percent, may be used. For example, for a depth of discharge of 70 percent the service life is about 1080 cycles.

Data on the charge–discharge cycle life of nickel–cadmium batteries is costly to acquire. The U.S. Navy's Naval Weapon Support Center at Crane, Indiana, has been testing nickel–cadmium cells for space use for decades. Each test requires 10 to 50 cells because at each temperature there must be a statistically significant number of samples. Sometimes the life is long, for instance, 10 to 25 years, and technology that produced the successful cells changes during the years of testing.

Reliability of Nickel–Cadmium Batteries An example illustrates the reliability of nickel–cadmium batteries for spacecraft. Traditionally, redundant batteries have been used to avoid loss of spacecraft functioning if a cell fails. A lightweight mechanization of redundancy is to add an extra cell or cells in series in the battery so that if one cell fails shorted, the battery will still produce minimum voltage. Some batteries have diodes to bypass cells that fail in the open-circuit mode.

Thomas Dawson of the U.S. Air Force Space Division, in planning future FLTSATCOM satellites, wondered if this redundancy is necessary. He interviewed project engineers responsible for six spacecraft and found that during 107.2 satellite years of operation there had been no open-circuit failures [9]. One satellite had experienced four cell shorts in 33 satellite years. Two others had experienced no battery cell failures in 53 years of satellite operation, with the longest satellite life exceeding 8 years.

Major Dawson concluded that rather than installing bypass diodes, the money and weight could better go into extra battery weight that reduces the stress on the batteries and increases their life. Eliminating the battery cell bypass electronics also increased the reliability of the battery assemblies from 0.9691 to 0.998.

Nickel Battery Cost   Nickel–cadmium batteries cost more than lead–acid batteries. For example, nickel costs around $3 a pound. Cadmium, a by-product of zinc refining, costs $1.50 per pound. Lead is around 25 cents per pound. In seeking a lower cost battery for electric automobiles, M. Klein and A. Charkey built nickel–cadmium batteries for electric vehicles by using the technique that had been developed for the miner's lamp battery [10].

Klein's nickel–cadmium battery powered a Saab with a curb weight of 2550 lb. The travel range was 75 miles. With an improved battery, the car would weigh 2180 lb and have a range of 100 miles. However, Klein saw the electric car becoming competitive with an internal combustion engine car making 27.5 mpg only when the price of gasoline reaches $2.45 a gallon.

3.3.4   Gas Metal and Other Spacecraft Batteries

The nickel–hydrogen battery is a successful gas–metal battery that is now being used in COMSAT synchronous-orbit satellites. The nickel–hydrogen cell is like a nickel–cadmium cell, except that the cadmium electrode is replaced by a platinum electrode that consumes hydrogen gas during discharge and evolves hydrogen when being charged. The hydrogen pressure can reach 200 kg/cm^2 when the cell is fully charged. Hughes Aircraft Company, using a cell design developed for the U.S. Air Force, has tested 25-Ah nickel–hydrogen cells at 80 percent depth of discharge for 7000 charge–discharge cycles. The observed capacity at that point, reported by E. Levy, was in the range of 23 to 30 Ah [11]. A nickel–cadmium cell at 80 percent depth of discharge would last only around 1000 charge–discharge cycles.

An advanced nickel–cadmium battery might deliver 55 Wh/kg at 100 percent depth of discharge. A nickel–hydrogen battery can deliver 65 Wh/kg. Recent cells have a life goal of 30,000 charge–discharge cycles [12]. Although rechargeable nickel–hydrogen batteries are good for spacecraft, they are not necessarily useful for terrestrial uses. They do require a platinum catalyst, which adds to their cost.

Redox Flow Battery Stores Energy in Liquids   The redox flow battery, developed at NASA Lewis Research Center, stores energy in electrolytes from which the energy is extracted with carbon electrodes. The electrolytes can be stored in tanks, which can be made as large as needed for the energy to be stored.

K. Nozaki and his colleagues developed improvements and built a 1-kW 30-cell unit that they tested. With computer simulations based on test data they predicted an 80 percent in-and-out efficiency for a 1-MW unit that they planned to build [13]. In selecting electrode materials they had screened 100 varieties of carbon before adopting carbon that is knit from cellulose. No catalyst was required. The positive electrolyte was iron chloride dissolved in hydrochloric acid. The negative electrolyte was chromium chloride, also dissolved in hydrochloric acid.

H. L. Steele and L. Wein of the Jet Propulsion Laboratory list the redox battery as the potentially lowest cost mechanism for storing energy in electrochemicals [12]. They estimate that the initial cost of a redox battery will be 20 percent of a lead–acid battery's cost.

3.3.5 Lithium Ion Batteries Deliver Long Life with Low Weight in Electric Bicycle Propulsion

The electrolyte concentration changes that occur during charging and discharging in nickel–cadmium batteries are similar to those in lead–acid batteries. However, a lithium ion cell contains no liquid electrolyte. In it a sheet of lithium is clamped between the cobalt oxide positive plate and the graphite negative plate. Charging converts the positive plate surface to lithium cobalt oxide and the negative plate surface to lithium carbonate. As a result, there is no electrolyte leakage problem, and every cell can be permanently sealed during its manufacture. Also, lithium ion cells need not be oriented with respect to the direction of gravity in vehicles or other applications.

The rugged lithium ion batteries successfully propelled the Mars rovers Spirit and Opportunity as they explored the planet after having survived complex stresses that started with the intense vibration during their launch from Earth. The batteries were stored with little discharge occurring during the zero-gravity coast from Earth to Mars. Intense deceleration and vibration of rocket firing for insertion into Mars orbit followed. Further deceleration came during the descent in the Mars Lander. Then on the planet's surface the batteries were recharged with solar power. The lithium-ion-battery-powered Spirit and Opportunity then roamed on the Mars surface to help solve scientific mysteries.

Alternative Configurations of Lithium Ion Batteries Being mass produced are type-18650 lithium ion cells that power communication equipment. Each cell is 18 mm (0.71 inches) in diameter and 65.5 mm (2.58 inches) long. A cell's mass is typically about 13.5 g. Consequently, high-power lithium ion battery development has progressed in two directions. The first involved assembling these type-18650 lithium ion cells into series strings that are connected into parallel modules that produce the required current and have the required ampere-hour energy content. These modules are then connected in series to form the battery that delivers the required voltage and energy with a depth of discharge that results in the required battery life. The advantages of this approach are (1) the mass-produced type-18650 cells have high proven reliability, (2) lifetimes up to 50,000 charge–discharge cycles have been demonstrated, and (3) the cost of cells is decreasing as more firms start producing them. However, a battery of this type requires sophisticated internal controls to prevent any possible single failure from disabling the battery.

The second battery development direction is to build larger lithium ion cells so that the new battery will have fewer cells than has a lead–acid, nickel–cadmium, or nickel–metal hydride battery with the same terminal voltage. This requires designing and testing a new cell to confirm its performance in terms of charge–discharge cycles.

The key component of a useful electric bicycle is the battery in which energy is stored for supplementing pedal power. The available battery types range from lead–acid, which can deliver 20 Wh/kg of battery weight to new lithium ion batteries that can deliver 160 Wh/kg. Research laboratories have even built lithium cells that deliver up to 400 Wh/kg [3]. SAFT America has developed its type VLE cell that features a specific energy of 195 Wh/kg. AGM Batteries Limited has obtained 235 Wh/kg in its 6-Ah 36550 cells that are based on stabilized lithium metal powder (SLMP). In-cell electronic overcharge protection detects and deviates the charge current when the cell becomes fully charged [4].

Saft America developed a new high-power lithium cell in a cost share program with the Partnership for a New Generation of Vehicles (PNGV) and the U.S. Department of Energy. N. Raman described the results of this work [14]. The cell specifications for this very high power 4-Ah cell for operation at 25°C are summarized in Table 3.1. When discharged at 500 A into a load, it delivered 82 percent of its rated capacity and its temperature did not rise above 40°C. Tests showed that the cells can meet cold-cranking requirements at −30°C and can provide the required power even after 30 days of storage at 25°C.

Lithium Ion Battery Tested in Propelling an Electric Bicycle At the 41st Power Sources Conference in 2004 Stephen S. Eaves described a new 76-Ah lithium ion battery that is built from small commercial 18650 Li ion cells by using massively parallel modular architecture [15]. He illustrated his technology with the ModPack lithium ion, a rectangular battery with dimensions of about

TABLE 3.1 Characteristics of Lithium Ion SAFT[a] Cell for HEV

Case diameter, cm	3.4
Case length, cm	15.3
Mass, kg	0.32
Total energy, Wh	13.7
Available energy,[b] Wh	5.5

Source: From Raman et al. [14].

[a] SAFT America VLE series VL4V cell.

[b] 30 to 70 percent of state of charge.

30 cm long by 23 cm wide by 8.3 cm thick operating at 14.8 V, with a capacity of 76 Ah. Within this ModPack the lithium cells are connected in parallel and series to achieve a specified capacity and voltage. Finally multiple modules are connected in series to obtain the desired battery pack voltage. Embedded electronic circuitry reliably balances the capacities of the individual cells in the module, and a supervisory battery management system enables the user to properly interface the battery to his application bus. The modular control circuitry has the ability to disable a section of a module that contains a failed cell with only a minor loss in module capacity, thus providing a high level of reliability.

Using mass-produced commercial Li ion cells avoids the cost of buying bigger cells that are not yet in mass production. The discharge voltages of the ModPack during test discharges are plotted in Figures 3.8 and 3.9.

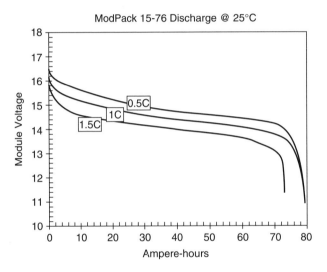

Figure 3.8 Output voltage of the Modpack battery when discharged at rates of 0.5 C, 1 C, and 1.5 C. (From Stein [17].)

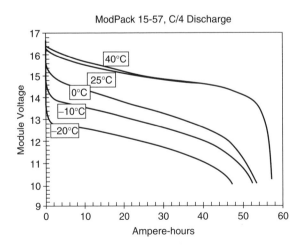

Figure 3.9 Output voltage of the Modpack battery at various temperatures when discharged at $C/4$. (From Stein [17].)

Testing a Lithium Ion ModPack-Powered Electric Bicycle An electric bicycle, manufactured by eGo Vehicles Corp. of Providence, Rhode Island, had been entered into a Tour De Sol competition organized by the Northeast Sustainable Energy Association. It was powered by a lead–acid battery and had a top speed of 32 km/h. In the contest its travel range was 32 km. Replacing the lead–acid battery with a ModPack Li ion battery reduced the bicycle's curb weight from 55 to 48 kg, raised its speed to 50 km/h, and more than doubled its travel range to 90 km. It consumed 21 Wh/km while traveling at speeds up to the 50 km/h over that 90-km distance on one battery charge. The cost of the electric energy for this 90-km (56-mi) trip was around 16 cents.

Details of the performance comparison are in Table 3.2. Not shown is the significant difference in the lifetime of the two batteries in charge–discharge service. The Li ion 18650 cells that were built into the ModPack had survived over 20,000 deep charge–discharge cycles in previous tests. Li ion 18650 cells had also been assembled into the battery that propelled the tZero two-passenger electric car, which consumed 33.8 kWh of electric energy in traveling 153 miles (246 km). Details of the tZero's performance are described in Section 3.2.3.

Complex problems, such as the selection of a battery type for an electric bicycle, are best solved by an aerospace decision-making procedure called system engineering, which is described in Chapter 6. The first step in this procedure is to quantify all of the requirements of the system that is being developed. The second step is to create alternative designs that fulfill these requirements with the required performance and reliability, and quantify risks identified for each design. The third step is calculating the life-cycle cost of each design and quantifying all of the risks involved. The best option, with risks cited, is then offered to the decision-making authority.

Estimating Lithium Cell Capacity After 35,000 Charge–Discharge Cycles
How would a lithium ion battery perform after 7 years in low Earth orbit? The European Space Agency had planned in 1999 the use of lithium ion batteries in such a spacecraft, so life tests were started in May, 1999, on six 100-Ah cells manufactured by Mitsubishi Electric Corporation. The specification of this battery is summarized in Table 3.3. The cells being tested had passed 35,000

TABLE 3.2 eGo Bicycle with Lead–Acid and Li ion Batteries

Model	eGo Cycle 2	eGo XR
Curb weight, kg	54.5	46.8
Battery weight, kg	22.7	15.1
Battery energy, Wh	720	1900
Nominal voltage, V	24	30
Top speed, km/h	40	50
Range, km	32	90

Source: From Stein et al. [17].

TABLE 3.3   Specifications of LEO 100 Ah Li Ion Cell

Parameter	Specification
Shape	Elliptic cylindrical
Dimensions, $H \times W \times T$, cm	20.8, 13, 5
Mass, kg	2.79
Case material	Aluminum alloy
Positive electrode	Lithium cobalt dioxide
Negative electrode	Carbon
Separator	Microporous film
Electrolyte	Li salt[a]
Nominal discharge voltage	3.7
Specific energy, Wh/kg	133

Source: From Inoue et al. [16].
[a]Li salt dissolved in mixture of alkyl carbonate solvents.

charge–discharge with a 30 percent depth of discharge in each cycle. Take-fumi Inoue summarized the acquired data at the Second International Energy Conversion Engineering Conference in 2003 [16].

The capacities of each cell were measured at a 20°C temperature before the life test and periodically during the life test. Each capacity measurement was preceded by a 25-A constant-current charge, followed by 3.98-V constant-voltage charge. Each cell's capacity was first measured at a 20-A ($C/5$) current, and then after recharge at a 50-A ($C/2$) current.

Battery 4 was discharged at 60 A for 30 min, left at open circuit for 10 s, and then recharged at 25 A current. This was followed initially by 3.95 V charging for 85 min. The voltage was raised to 3.98 V after 16,000 cycles. Battery temperature was kept at 20°C. Inoue's plots of the end-of-discharge voltages in battery 4 after 35,000 of charge–discharge cycles are shown in Figure 3.10. The voltages of three representative cells had declined from 3.7 to 3.55 V during the 35,000 cycles of charge–discharge testing.

Mitsubishi had developed these cells for a 40-kWh battery that could be used on hybrid electric vehicles. The requirements included a useful life of 300,000 charge–discharge cycles, each having 25-kW power pulses. The battery cost goal was $500 per battery.

Future of Lithium Ion Batteries   The long lifetime and low weight of lithium ion batteries makes them the best choice for many energy storage applications. However, their high cost does limit their marketability. Their cost is bound to decline as new technology is applied into producing the rare-earth elements required in these cells.

In electric bicycle propulsion the basic requirement is low weight and low usable-lifetime cost. Lithium ion batteries excel in these respects. Fuel costs are going to rise after the world's petroleum production peaks, and users will then be willing to pay the high price of a bicycle in which the battery that stores propulsion energy has decades of predicted service life.

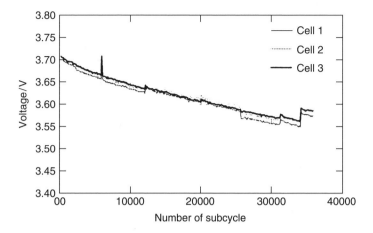

Figure 3.10   During 35,000 charge–discharge cycles the end-of-discharge voltage dropped from 3.7 V per cell to 3.55 V. (From Inoue et al. [16].)

Lightweight Lithium Ion Cells in Aluminum Case   Cells built for aerospace application must meet requirements that include intense vibration, acceleration, shock, and temperature, plus the ability to carry overloads in emergencies. Brian J. Stein described progress at Mine Safety Appliances Company in developing a lithium ion cell in a case that meets these requirements [17]. This led to consideration of aluminum and titanium as lightweight materials with densities of 2.70 and 4.51 g/cm^3. Stainless steel weighs 8.03 g/cm^3. Replacing stainless steel in a 10-Ah cell would result in weight savings of 13 percent and a 15 percent increase in gravimetric energy density of the cells.

The critical component of a cell assembly is its case. Case alloys evaluated include titanium and aluminum alloys 1100, 3003, and 6061. Alloy 3003 was chosen because of its high purity of 98.20 percent minimum and because it has 1.0 to 1.5 percent magnesium which gives it 20 percent extra strength. A thin-walled case with exacting dimensions is needed to properly constrain the electrode stack and provide consistent interelectrode spacing throughout the electrode stack during cycling. The case could have been made as a two-piece fabricated/welded assembly. A deep-drawn case was chosen because of its high reliability. The integrity of a unit structure is repeatedly consistent in dimensions, and the unit structure is lower in cost than that of a welded structure. Testing showed that the aluminum alloy case met corrosion requirements. Safety tests included vibration, short circuit, and overcharge effects.

The resulting prototype cell is shown in Figure 3.11. Charge–discharge cycling test results, on 10- and 50-Ah cells, are plotted in Figure 3.12. They show that at 40 percent depth of discharge, the 10- and 50-Ah cells had acceptably low loss in capacity after 10,000 charge–discharge cycles. The authors concluded that this program produced a lighter lithium ion cell with all of the mechanical and electrical performance characteristics of small cells assembled into stainless steel hardware.

Figure 3.11   Lightweight aluminum housing adds ruggedness to a 10-Ah lithium ion cell. (From Brian J. Stein [17].) Reproduced with permission of Mine Safety Appliances Company.

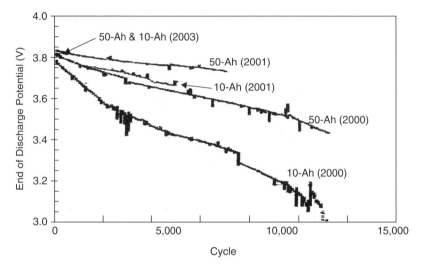

Figure 3.12   New 10- and 50-Ah lithium ion cells offer a 10-year life in daily trips on an electric bicycle. (From Brian J. Stein [17].) Reproduced with permission of Mine Safety Appliances Company.

3.3.6   Zinc, Sodium, and Other Zinc Batteries

Zinc is a high-energy low-cost electrode in primary batteries, such as the common dry cell. Zinc has also been a successful electrode in silver–zinc primary batteries that deliver over 100 Wh/lb. These batteries can be designed to deliver their energy in a few minutes, making them useful for missiles, launch vehicles, and torpedoes. Such high-discharge-rate batteries are often constructed so that the electrolyte is introduced just before use.

Development of rechargeable silver–zinc and nickel–zinc batteries has been only partly successful. During charge the zinc is plated out of solution, and it tends to form dendrites through pinholes in the material separating the plates. The pinholes seem to provide a low-resistance path that attracts dendrite growth. Some silver–zinc batteries have achieved over 100 charge–discharge cycles. Nickel–zinc batteries have survived 700 cycles but not consistently.

Sodium Sulfur Batteries for 100 percent Coulombic Efficiency   Sodium sulfur batteries have been developed for electric cars. Their advantage is their high power density, which is around 110 Wh/kg. A disadvantage is that their operating temperature is 350°F, which requires an oven.

Douglas M. Allen decided to test these batteries for possible use in spacecraft, where an excellent vacuum insulation is available for the oven. The test results were dramatic. By August 4, 1984, the number of charge–discharge cycles at 80 percent depth of discharge had reached 2400, with no discernible loss of performance [18]. The charge voltage was higher than discharge voltage, so the overall efficiency was 80 percent. Open-circuit voltage of these cells is 2.07 V.

U.S. Department of Energy's Battery Development for Electric and Hybrid Vehicles The U.S. Department of Energy's (DOE) response to the coming petroleum crisis is a new program, Energy Efficiency and Renewable Energy. This program was described at plenary session, "Advanced Energy Storage Technologies—Potential Applications and Status" at the First International Energy Conversion Engineering Conference in 2003. The program's objective, described by James A. Barnes in an oral presentation, is to make battery-powered and hybrid electric vehicles practical [19].

Major challenges of this program cited by Barnes are cost, abuse tolerance, and battery life. Among commercial long-term goals of the battery that powers 40-kW electric vehicles in this program were the following:

Specific power during discharge to 80 percent depth:	400 Wh/kg
Cycle life with 80 percent depth of discharge	1000 cycles
Cycle life with 30 percent depth of discharge	2670 cycles
Normal recharge time	3 to 6 h
High-rate charge, from 40 to 80 percent state of charge	15 min

3.4 FUEL CELLS FOR POWERING ELECTRIC BICYCLES

Energy conversion efficiencies of over 90 percent have been achieved in hydrogen–oxygen fuel cells. The benefit of this efficiency was illustrated by a Brazilian author who described a solution to the problem that their utilities have in supplying peak power during late afternoons. Residents of apartments and homes could help supply power during this peak power period from fuel cells. During night times they can electrolyze water into hydrogen and oxygen for their fuel cells. The utility could reward these helpers because they would avoid the cost of constructing a new power plant that runs only during peak load periods.

Fuel cells that consume compressed or liquefied gases have efficiently generated power for manned spacecraft. Fuel cells have not been practical for Earth surface power generation because storing and transporting hydrogen is difficult, and the required platinum catalyst is costly. Now new technology is making fuel cells practical for powering electric vehicles. For example, hydrogen that can be extracted from natural gas could now be stored compactly in hydrides. Fuel cells can also run on liquid fuels. Prototype fuel-cell-powered buses are in operation.

The high efficiency that is possible with fuel cells could make fuel-cell-powered bicycles practical for traveling long distances, as well as for local travel, during the coming postpetroleum era. In the sections that follow we review the fundamentals of fuel cells and how they might be used to propel electric bicycles. Present-day problems are the high cost of hydrogen fuel and the difficulty of carrying it on a bicycle. Possible solutions for these problems are evaluated.

3.4.1   Fuel Cell Electric-Power-Producing Basics

A fuel cell is basically an electrochemical cell, like a battery cell, that consists of two plates that are separated by an electrolyte. The negative electrode, which contributes electrons to the external circuit, is also called an *anode* during discharge. The reaction at the negative electrode is called *oxidation*. The positive electrode or cathode absorbs electrons from the external load. The reaction at the cathode is called *reduction*. A memory help is "red cat" for the reduction that occurs at the cathode.

A typical fuel cell is a box with catalyzed electrodes, gas passages, coolant passages, and electrolyte that converts the chemical energy of incoming gases, usually hydrogen and oxygen, to electric power. Usually, the fuel cell is more efficient and compact if the reaction takes place under a pressure of several atmospheres and the temperature is above the boiling point of water. The basics of a proton–electron membrane (PEM) hydrogen fuel cell are summarized in Figure 3.13. Hydrogen enters the dry side of the anode plate, and its electrons are stripped off of the hydrogen atoms before they enter the electrolyte. The electrons flow through the load to the cathode of the cell. The hydrogen ions, after leaving their electrons at the anode, flow through the electrolyte to the

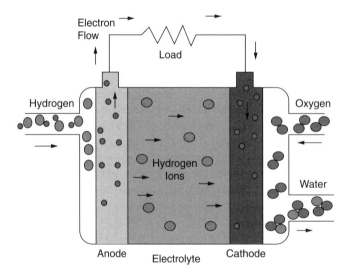

Figure 3.13   In a fuel cell the hydrogen electrons separate from their ions at the negative anode, then flow through the external load, and finally rejoin their ions at the cathode to form water. (From S. Johnny Fu, Ansoft Corporation from presentation made at Boeing, Kent, WA "Workshop, 2002," p. 36.)

cathode, where they reunite with their electrons and combine with oxygen gas to form water that flows out of the cell.

The reactants that enter the fuel cell are called *fuel* and *oxidizer*. The fuel is usually hydrogen, coming from a tank, hydride bed, or from a reformer that converts methanol or methane into hydrogen. The oxidizer is usually pure oxygen or air. The output volt/ampere characteristics of a typical fuel cell are plotted in Figure 3.14. This "PMFC" curve is for a cell with a typical proton–electron membrane that separates the anode from the cathode. Most fuel cells are built around proton–electron exchange membranes.

Energy features of fuel cells include the following:

- Conversion of fuel and oxidizer into electric power is not a heat engine process, so Carnot efficiency limits do not apply.
- Fuel cells have been successful power sources for manned spacecraft where power is needed during boost and reentry mission phases, as well as during cruise. High-purity hydrogen is needed for the fuel in spacecraft fuel cells.
- Fuel cells have supplied commercial and utility power, but the economic advantages of fuel cells over alternative power sources have not been proven.
- Platinum-type catalysts are generally required in fuel cells.
- Hydrogen is hard to store, so terrestrial fuel cells are often designed to use methane or liquid fuels. The required reformers would add to the complexity and losses in the power plant. Hydrogen can also be stored in liquid ammonia, and a newly discovered catalyst could dissociate ammonia to release hydrogen gas for a fuel cell that propels an electric bicycle.

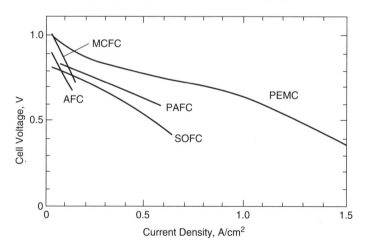

Figure 3.14   Efficiency of a fuel cell, when carrying a given load current, is the voltage at load current divided by the voltage at zero current. At high currents the efficiency is low. AFC is acid fuel cell, PEMFC is proton exchange fuel cell, PAFC is phosphoric acid fuel cell, SOFC is solid oxide fuel cell, and MCFC is molten carbonate fuel cell. (From S. Johnny Fu, Ansoft Corporation from presentation made at Boeing, Kent, WA, "Workshop 2002," p. 34.)

An inherent energy loss of at least 17 percent occurs when hydrogen and oxygen are converted into water.

The key performance parameters of fuel cells are voltage and current density. The cost of a fuel cell is related to the electrode area, so designers look for high current density. Every fuel cell reaction has a theoretical voltage and an achievable voltage. The difference from these voltages represents losses, which come from fuel. The fuel cell designer's task is to find the best compromise of costly catalyst, clever cell configuration, operating temperature, and sophisticated auxiliaries.

The performance of a single cell is usually plotted in terms of voltages as a function of amperes/square foot, or of milliamperes/square centimeter. Efficiency is the ratio of power output of the cell to the heating value of the fuel used to produce this power.

3.4.2   Fuel Cell Performance

Fuel cell performance is measured in heat rate, expressed as the higher heating value of the fuel consumed in generating 1 kWh of electricity. A 100 percent efficient fuel cell would consume only 3412 Btu/kWh of fuel. The United Technology 4.8-MW plant at Tokyo Electric had a fuel rate of 9600 Btu/kWh. The best U.S. steam plant in 1983 had a fuel rate of 8987 Btu/kWh. In 2003 the best combined-cycle steam plant had an efficiency of 60 percent, and its fuel consumption was only 5690 Btu/kWh. Its construction cost was $500/kW of power output.

The electrochemical performance of a fuel cell is plotted in volts as a function of current density, commonly in amperes/square feet. The energy released when hydrogen and oxygen combine in combustion corresponds to 1.48 V output of a fuel cell if water is the product and 1.23 V if it is steam. Neither voltage can be achieved because of the irreversibility of the electrode processes, activation polarization, and concentration or activity gradient in the electrodes.

An objective of fuel cell developers is to get the highest practical voltage from the cell. A. J. Appleby, of the Electric Power Research Institute, in a paper presented in a fuel cell session at Orlando, Florida, developed a value for this voltage [20]. He assumed that the annual fixed cost of a fuel cell is $200/kW per year, and that the plant would have a 5000 hours-per-year availability. If the fuel, natural gas, costs $7/million Btu, then a 40-mV improvement gives the same overall electricity cost change as a 7.5 percent decrease in capital cost of the power plant.

High Cost of Transporting and Delivering Hydrogen   Hydrogen supply problems have been quantified but not yet solved. At 11,600 psi the energy in a given volume of hydrogen is 10.2 megajoules per liter (MJ/L) as shown in Table 3.4. That is only one-third of the energy in gasoline or in methane at 11,600 psi.

Natural gas is produced from drilled wells and is compressed at the wellhead for delivery to consumers in a pipeline. The pipe has the cross-section area and wall thickness that enables it to deliver the required quantity of energy to its

TABLE 3.4   Energy Density of Various Fuels

Fuel	State	Pressure (psi)	Energy Density (MJ/Li)
Gasoline	Liquid	Ambient	33.5
Methane	Gas	11,600	32.5
Propane	Liquid	Ambient	25.2
Methanol	Liquid	Ambient	17.5
Hydrogen	Gas	11,600	10.2
Hydrogen	Gas	2,900	8
Hydrogen	Liquid	~200 [a]	10
Methane	Gas	2900	2.5

Source: John R. Wilson, Why (Not) Hydrogen? Electric Vehicle Association, Current Events, Nov–Dec 2003.

[a] At a temperature of $-240°$C.

gas-receiving destination. Assume that the compressor's output is changed to hydrogen from a processor, and pipeline's pressure limit is not exceeded. Then the energy delivered to the using agency in the hydrogen would be only one-third of the energy in the previously delivered natural gas. Also, much carbon would need to be shipped out of the wellhead location where the hydrogen is extracted from the methane and other hydrocarbons in natural gas.

Today's lowest cost hydrogen is a by-product of oil refineries. The world's petroleum production is expected to peak in the year 2005, and diminish every year after that. Thus the cost of hydrogen derived from petroleum refining can be expected to rise. Hydrogen can be produced at gasoline filling stations by electrolyzing water. An ultimate source of hydrogen could be a high-temperature electrochemical process that uses the heat energy from coal to create a hydrogen-producing reaction that also releases much carbon dioxide.

Solar energy can also be used to produce hydrogen. A solar array with 34 percent efficient solar cells could deliver its output energy to an electrolyzer that can produces hydrogen with an efficiency of over 80 percent. Plant leaves use the energy from 3-eV photons for producing hydrogen that they convert into hydrocarbons. However, less than 30 percent of the solar spectrum has enough energy for directly extracting hydrogen from water. Solar cells do capture the energy even from the green light that plant leaves reflect away.

Nuclear power plants are the obvious source of energy for electrolyzing water for hydrogen production. Nations, such as China, are now building more nuclear power plants.

3.4.3   Fuel Cell Power for Electric Bicycles

Liquefied hydrogen and oxygen are carried in insulated tanks in manned spacecraft where they are combined in fuel cells to generate the needed electric power. Heat leaking into the liquefied-gas tanks would cause the gases to boil and generate high pressure, so the hydrogen and oxygen tanks are stored in thermally insulated containers. Multilayers of reflective insulation separated the liquid-gas-containing tanks from the warm environment in the spacecraft. To avoid heat transfer by convection air currents, the layers of reflective insulation have to be evacuated. A vacuum is readily available on a spacecraft but not conveniently available on a bicycle.

The fuel-cell-powered Opel Zafira, a five-passenger minivan, was the marathon pace car that had been built for the 2000 Olympics. Its propulsion power from a 75-hp ac motor enabled it to accelerate from 0 to 60 mph (96 km/h) in 16 s. Its top speed was 84 mph (135 km/h). An air compressor supplied oxygen to the fuel cell stack. Its hydrogen supply was extracted by a reformer from methanol and then stored as a liquid in a tank at a temperature of $-253°C$. The stainless-steel tank that can store 2.27 lb (5 kg) of liquid hydrogen is 1 ft (30 cm) in diameter and 3.25 ft (1 m) in length. Multilayers of fiberglass insulation surround the tank. A high-voltage battery pack under the Zafira's rear floor supplements the fuel cell power when peak power for acceleration is required.

3.5 BEST NEW ELECTRIC POWER SOURCES FOR BICYCLE PROPULSION

In 1999 we powered with lead–acid batteries the electric bicycles that we used for measuring the power required for electric bicycle travel over streets, highways, and over high hills. The lead–acid battery was our most practical propulsion energy source at that time. Today's rapidly growing electric-powered vehicle needs have produced development of new and better propulsion power sources that are described in the sections that follow.

3.5.1 Zinc–Air Fuel Cell Power for Electric Bicycles

An electric bicycle that is propelled by a fuel cell that runs on hydrogen from a tank and oxygen from the air could have an efficiency that permits traveling 1000 miles by consuming the energy equivalent of 1 gal of gasoline. However, the hydrogen in either a high-pressure tank or cryogenic-liquid container can be difficult to carry on a bicycle.

A recently developed fuel cell that runs on powdered zinc, plus oxygen from the air, can become a power source for electric bicycles. It has been tested in powering buses that ran hundreds of miles through the Alps in Europe. The zinc powder is fed into the cell until the cell is fully loaded with zinc oxide (Fig. 3.15). Recharging requires pumping out the zinc-oxide-loaded electrolyte and replacing it with fresh electrolyte. The zinc is then extracted from the zinc oxide by electrolysis, which could even be powered by a solar cell array. In a remote region a container of zinc powder could be kept on hand for powering the bicycle on cloudy days.

In another design the zinc pellets are dropped into the top of the zinc–air fuel cell, between the negative and positive plates, as shown in Figure 3.16. Electrolyte is injected into the bottom of the cell and circulates out at the top of

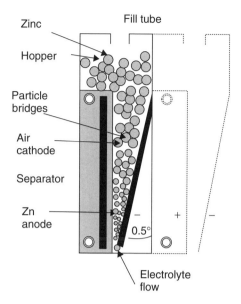

Zinc

Hopper

Particle bridges

Air cathode

Separator

Zn anode

Fill tube

0.5°

Electrolyte flow

Figure 3.15   Electrolyte-containing zinc–air fuel cell receives zinc pellets from a hopper and oxygen from air that is delivered through the cathode by a blower. (From Stuart Smedley, Metallic Power Inc., Carlsbad, CA, "Zinc Air Fuel Cell for Industrial and Specialty Vehicles," December 2000.)

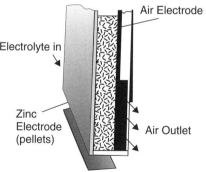

Zinc Pellets

Negative Terminal

Positive Terminal

Zinc Pellet Hopper

Air Inlet from Air Blower

Electrolyte Out

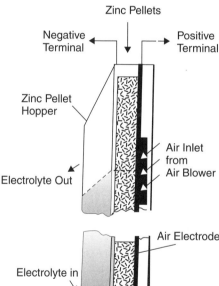

Air Electrode

Electrolyte in

Zinc Electrode (pellets)

Air Outlet

Figure 3.16   In this zinc–air fuel cell the electrolyte flows upward and clears the cell area of the zinc oxide, which forms as power is generated in the active zone in the fuel cell. (From N. J. Cherepy, R. Krueger, and J. F. Copper, A Zinc/Air Fuel Cell for Electric Vehicles, Lawrence Livermore National Laboratory, Livermore, CA.)

the cell. Oxygen is supplied in the air that is blown past the outside of the positive plate, which is carbon through which oxygen can flow. In one version of the cell the zinc pellets are fed into a hopper and drift downward in the electrolyte as they lose zinc into the electrolyte.

Metallic power has developed and tested the vehicle-refueling station shown in Figure 3.17. The zincate-loaded electrolyte is pumped out from the vehicle's fuel cell assembly and routed into the "sprouted bed electrolyzer." Other hoses at this refuel station deliver fresh electrolyte and zinc pellets into the vehicle's tanks. The vehicle then leaves the station, and the electrolyzer proceeds to separate the zinc from the electrolyte, using local electric power in the process that recovers the zinc.

Zinc–air Cell Electrochemistry A common anode for primary flashlight and other cells is zinc, a plentiful metal that costs around 50 cents/lb. It provides a convenient 1.2 to 1.7 V per cell output as its zinc is oxidized into zinc oxide or zinc chloride.

The zinc–oxygen reaction is useful because only the anode, the electrolyte, and a nonreacting carbon–cathode are required. This is the reaction in the "air cell" battery used in early vacuum tube radios. It is also used in the zinc–air battery that is manufactured by McGraw Edison for railroad signaling, ice floe beacons, radio repeaters, and offshore buoys. The buoys are hard to service, and

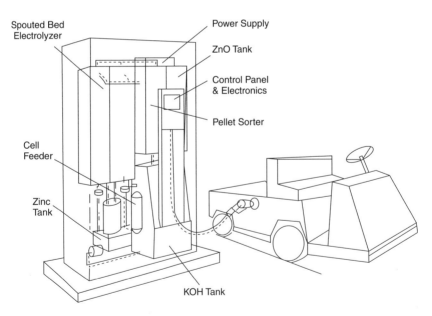

Figure 3.17 The "refuel station" sucks out the zincate-containing electrolyte in the fuel-cell-powered vehicle. It then replenishes the vehicle's tanks with fresh electrolyte and zinc powder. The "spouted bed" electrolyzer uses utility power to recover the zinc from the received liquid. (From Stuart Smedley, A Regenerative Zinc Air Fuel Cell for Industrial and Specialty Vehicles, IEEE 15th Annual Battery Conference on Applications and Advances.)

the need to recharge a storage battery would add to the complexity of the task. The zinc–air battery can be discarded after use. At a buoy, where a lead–acid storage battery of given weight has to be recharged every 6 months, the zinc–air battery of the same weight will deliver power for 5 years.

The zinc–air battery uses a porous carbon electrode for delivering oxygen from the air into the electrolyte (Fig. 3.16). Here oxygen reacts with water, receiving electrons from the external circuit, to form hydroxyl ions:

$$O_2 + H_2 + 2e^- \rightarrow 2OH^-$$

The hydroxyl ions diffuse through the electrolyte to the zinc anode for this oxidation reaction:

$$Zn + 4OH \rightarrow ZnO_2 + 2H_2O + 2e^-$$

At the negative electrode four OH^- ions are consumed for every two generated at the positive electrode. The released electrons travel through the external circuit to the carbon cathode, delivering energy to the load. The zinc consumption is related to the delivered current. One gram mole of zinc is its atomic weight, 65.38, expressed in grams. One gram-mole of an element has 6.0225×10^{23} atoms, and one electron is worth 1.602×10^{-19} coulombs. With these constants we calculate the consumption, C_z, of zinc in a battery, noting that each atom contributes two electrons to the reaction.

$$C_z = \frac{2 \times 6.02252 \times 10^{23} \text{ electrons} \times 1.602 \times 10^{-19} \text{ coulombs} \times \text{mole}}{\text{mole} \times \text{electron} \times 65.38 \text{ g}}$$

$$= 2951.57 \text{ C/g}$$

One ampere-hour is 3600 ampere-seconds, so the gram of zinc gives:

$$\frac{2951.57 \text{ C/g}}{3600 \text{ C/Ah}} = 0.82 \text{ Ah/g}$$

This corresponds to 820 Ah/kg (372 Ah/lb), and at 1.2 V/cell, 984 Wh/kg (446 Wh/lb).

Gas–Zinc Battery   The zinc–air battery produces about 1 Wh for every gram of zinc consumed. The reaction of zinc with chlorine and bromine releases more energy, hence gives higher voltage than those cited above. Such a battery is also called a hybrid fuel cell because one electrode is gas and other is metal. We once evaluated such batteries for an application needing many megawatt hours of stored energy. These electrochemical couples had already been developed for possible secondary batteries in electric automobiles.

Marti Klein of Energy Research Corporation developed for us concepts of a zinc–bromine battery. One of his concepts has a zinc-slab electrode that is consumed as the battery delivers energy. In a zinc-slab battery the electrolyte space widens as the zinc is reacted and carried away in solution. This increases the internal resistance of the battery and hence its internal losses. An alternative suggested by Klein is to deliver the zinc to the battery in the form of a slurry. Bromine is easy to store, being a liquid at temperatures up to 138°F at normal atmospheric pressure. The zinc–bromine cells would self-discharge, so the battery would have to be used within a few weeks after it is activated.

A zinc chlorine–gas primary battery was adapted for us from an electric automobile development by Energy Development Associates in Madison Heights, Michigan. Its battery had the zinc in a comb-shaped cross section. Carbon electrodes in the spaces between comb teeth were constructed with chlorine passages. The energy content of the battery was limited to its zinc content. The same principle is used in the firm's 500-kWh load-leveling battery for utility use.

3.5.2 Lithium Batteries

Lithium, a reactive lightweight metal, has been successfully packaged into batteries that deliver over 200 Wh/lb (440 Wh/kg). The U.S. Air Force has sponsored development of large units for its applications, and a spin-off has been long-life batteries for watches, emergency radios, flashlights, telephones, and laptop computers.

The energy content that can be extracted from a lithium battery is inversely related to the amount of power being extracted from the battery. For example, an Altus AL 1700-500 cell will deliver 500 Ah at nearly constant 3.5 V if it is discharged for 500 h at 1 A. If the current is increased to 6.25 A, the cell delivers only 300 Ah with rapidly dropping voltage (Fig. 3.18).

A lithium battery has about one-fourth the energy density of TNT. Accidents in the early development of lithium batteries motivated research that led to fail-safe batteries. For example, Altos has tested its lithium thionyl-chloride cells in

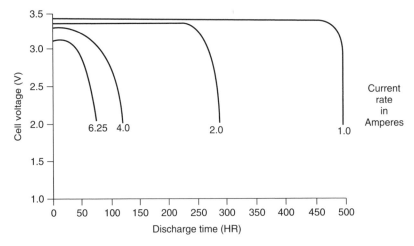

Figure 3.18   This lithium ion battery, when discharged at a 1.0-A rate, can deliver 500 Ah. When discharged at a 2.0-A rate its delivery drops to 300 Ah. (From H. Oman, *Energy Systems Engineering Handbook*, Prentice Hall, Englewood Cliffs, NJ, 1986.)

every conceivable accident scenario, including shock, vibration, puncture, crush, fire, reverse voltage, and forced charging. No catastrophes resulted.

A price must be paid for the light weight of lithium batteries. For example, an AL 1700–2000 battery, good for 200 Wh/lb, costs from $2500 to $5000, depending on quantity and configuration. At $2500, the price is $320/kWh. A zinc–air cell costs about $15/kWh.

3.5.3   Fuel Cell Power Plant Technology Can Be a Power Source for Electric Vehicles

The efficiency of fuel cells is increasing and the costs decreasing. Because fuel cells are not limited by the Carnot heat-cycle efficiency to around 60 percent, eventually fuel cell efficiency will exceed that of conventional power plants. The advantage of a fuel cell plant is that it can be bought in small increments that can be installed near loads and operated automatically. This introduces the possibility of a small business and even residences generating their own power. Using this power as a source of charging power for electric vehicle batteries could ultimately reduce airborne pollution as well as operating cost.

Cost of Fuel Cell Power Plants   Postulated costs of fuel cell power plants have ranged from $600 to $1000/kW. Meaningful costs will become available when plants are produced in a reasonable quantity.

The fuel cells being developed for public utilities use natural gas, methanol, and petroleum products for fuels. These fuels cost 3 to 5 times as much as coal on an energy basis. Thus the coal-burning steam plant is a logical alternative to fuel cells. The coal plant that is only 38 percent efficient can beat a hydrogen-consuming fuel cell plant that is 50 percent efficient, from a fuel cost viewpoint. The advantage of a fuel cell plant is that it can be bought in small increments that can be installed near loads and operated automatically. Also, the heat losses at a coal plant are not generally available to the customer. The business or residence(s) with the 40-kW natural gas burning fuel cell can get its hot water from the fuel cell's losses.

3.6   BICYCLE PROPULSION POWER SOURCES TO WATCH

The coming shortage of petroleum for fueling the world's transportation is motivating intense development of high-efficiency vehicles. For example, the petroleum production in the United States has already passed its peak, and currently the United States imports over half the petroleum that it consumes. Other nations are taking steps to reduce their petroleum consumption. For example, China's growing electric bicycle production has passed a rate of one million a year. The propulsion power is generated in the 18-GW Yangtze River hydroplant and in new nuclear power plants. A new factory is producing lithium ion cells for electric bicycles.

The intensity of development of new sources of power for vehicle propulsion is illustrated by the 547-page Proceedings of the 41st Power Sources

Conference that was held on June 14 to 17, 2004 (18). Authors of 147 technical papers described results from research and development programs that were searching for more efficient and practical technology for producing and storing energy, particularly in evaluating new concepts. For example, hydrogen is a fuel that can be efficiently converted into electric power for propelling vehicles. However, hydrogen is costly to store and transport to filling stations and inconvenient to carry on users' vehicles. Ammonia, which is composed of hydrogen and nitrogen atoms, is relatively easy to transport. One paper described a research program in which a catalyst was discovered for efficiently extracting the hydrogen from ammonia and delivering it to fuel cells. This discovery, if practical, might revolutionize motor vehicle transportation.

REFERENCES

1. Hosea W. Libbey, Electric Battery, U.S. Patent 536,689, April 2, 1895.
2. tZero Earns Highest Grade at 2003 Michelin Challenge Bibendum (cover story), *Electric Auto Association Current Events*, Vol. 35, Nos. 11 and 12, Nov–Dec 2003.
3. Frank B. Tudron et al., Lithium Sulfur Rechargeable Batteries: Characteristics, State of Development, and Applicability to Powering Portable Electronics, Proceedings of the 41st Power Sources Conference, 14–17 June, 2004, pp. 341–344.
4. Tony Jeffery and Jason Hinde, The Development of High Energy Density Lithium Ion Cells, Proceedings of the 41st Power Sources Conference, 14–17 June, 2004, pp. 438–441.
5. R. L. Proctor, Accurate and User Friendly State-of-Charge Instrumentation for Electric Vehicles, IEEE Northcon94 Conference, Seattle WA, November 4–6, 1994, pp. 379–384.
6. J. S. Enochs et al., Nonantimonial Lead Acid Batteries for Cycling Applications, Proceedings of the 19th Intersociety Energy Conversion Engineering Conference, ANS, 1984, pp. 850–856.
7. Willard R. Scott and Douglas W. Rusta, Sealed-Cell Nickel-Cadmium Battery Applications Manual, NASA Scientific and Technical Information Branch, Publication 1052, 1979.
8. William H. Lewis and Douglas W. Rusta, New Developments in Personal Lighting Systems for Miners, Information Circular 8938, U.S. Department of Interior, Bureau of Mines.
9. Thomas Dawson, Elimination of Battery Cell Bypass Electronics on FLTSATCOM, Proceedings of the 19th Intersociety Energy Conversion Engineering Conference, ANS, 1984, pp. 72–77.
10. M. Klein and A. Charkey, Nickel-Cadmium Battery System for Electric Vehicles, Proceedings of the 19th Intersociety Energy Conversion Engineering Conference, ANS, 1984, pp. 719–725.
11. E. Levy, Jr., Life, Engineering, and Acceptance Qualification Test Data on Air Force Design Nickel Hydrogen Batteries, Proceedings of the 19th Intersociety Energy Conversion Engineering Conference, ANS, 1984, pp. 85–88.
12. H. L. Steele and L. Wein, Comparison of Electrochemical and Thermal Storage for Hybrid Solar Power Plants, Paper 81-WA/Sol-27, ASME Winter Annual Meeting, 1981.
13. K. Nozaki et al., Performance of ETL New 1-kW Redox Flow Cell System, Proceedings of the 19th Intersociety Energy Conversion Engineering Conference, ANS, 1984, pp. 844–849.
14. N. Raman et al., Development of High Power Li-Ion Battery Technology for Hybrid Electric Vehicle (HEV) Applications. Proceedings of the 41st Power Sources Conference, 14–17 June, 2004, pp. 435–437.
15. Stephen S. Eaves, Large, Low Cost, Rapidly Configurable Lithium-Ion Battery Modules Constructed from Small Commercial Cells, Proceedings of the 41st Power Sources Conference, June 14–17, 2004, pp. 328–330.

16. T. Inoue et al., LEO Life Testing Results of 100 Ah Lithium Ion Cells, Proceedings of AIAA, pp. 2003–6018, Second International Energy Conversion Engineering Conference, August 2004, Providence, RI.

17. Brian J. Stein III, John W. Baker and Pinakin M. Shah, Development of Light Weight Lithium-Ion Cells Using Aluminum Hardware, Proceedings of AIAA, 2003–5987, First International Energy Conversion Engineering Conference, August 2003, Portsmouth, VA.

18. Douglas M. Allen, Sodium-Sulfur Satellite Batteries: Cell Test Results and Development Plans, Proceedings of the 19th Intersociety Energy Conversion Engineering Conference, ANS, 1984, pp. 163–168.

19. James H. Barnes, Presentation at Advanced Energy Storage Technologies—Potential and Status (Oral presentation at the first AIAA International Energy Conversion Engineering Conference, 2003).

20. A. J. Appleby, Acid Fuel Cell Technology—An Overview, Fuel Cell Seminar at Orlando, Florida, 1983.

CHAPTER *4*

BATTERY CHARGING

We tested a lead–acid battery to determine its lifetime by discharging it completely and then recharging it to a fully charged state in each charge–discharge cycle. At the end of 50 charge–discharge cycles we had a dead battery

    Sophisticated new battery-charging technology is key to the efficient and economic use of batteries for vehicle propulsion. In the sections that follow we (1) quantify for each battery the unique profiles in voltage, current, and depth of discharge that give long battery life and (2) discuss new automatic battery-charging controls that make long life possible.

4.1   HISTORY OF BATTERY-CHARGING TECHNOLOGY

Gaston Plante invented the lead–acid battery in 1859. By 1900 there were more electric cars than engine-powered cars in the United States. However, owners soon learned that buying a Model T Ford was much cheaper than replacing the failed battery in an electric vehicle. Then in 1912 Charles F. Kettering invented the lead–acid battery-powered electric starter for automobile engines. The starter was powered by a lead–acid battery. Subsequently brush-and-commutator generators on cars replaced the magnetos on cars. Automobile designers did not understand the requirements for long battery life, so even today lead–acid batteries for cars carry a life guarantee of around 60 months.

    The Cold War brought requirements for postattack power in missile silos and command posts. Then, after serious analysis, and even in testing in which one lead–acid battery exploded, we learned the limits of lead–acid battery technology. Then came the need to power Earth-orbiting satellites during their frequently occurring transits through the Earth's shadow when sunlight from solar cell arrays was not available. Years of intensive analysis and testing revealed that nickel–cadmium batteries could last for many years if not more than 30 percent of the battery's stored energy was consumed in each eclipse. The cost of launching these oversized nickel–cadmium batteries motivated intense battery development. Nickel–metal hydride batteries can now power Earth satellites for decades in Earth orbit.

Electric Bicycles: A Guide to Design and Use, by William C. Morchin and Henry Oman
Copyright © 2006 The Institute of Electrical and Electronics Engineers, Inc.

4.2 BASIC FUNCTIONS OF BATTERY CHARGERS

To understand the requirements of battery chargers, we review the performance characteristics of batteries that need charge control for achieving long life. The basic functions of battery charge control are:

- When supplied with public ac power, the bicycle's battery recharges to a full-charge condition after it has propelled the electric bicycle.
- Perform this recharge in a manner that preserves the battery's usable lifetime in propulsion service.

A voltmeter cannot show a true battery charge status because the terminal voltage in most batteries varies with load and battery temperature. Therefore, a charge control must integrate the current flow, during both charge and discharge, to track the battery's charge status for indicating discharge state. Charge controllers that use specially designed integrated circuits to perform these functions have been developed. These controllers are commercially available at reasonable prices.

The bicyclist needs to be aware of the state of charge of the battery on his bicycle so that he can avoid completely discharging the battery, which reduces battery life. For example, the Honda EV Plus carries a "management electronic unit" that displays to the driver the distance of travel available with the state of charge of the battery (Fig. 4.1). It also reports a "capacity reduction" factor that indicates aging of the battery. An ancillary function of battery charging would be to report to the cyclist, during his travel, the charge status of the battery, and warn him when the battery's completely discharged condition is approaching.

4.3 BATTERY CHARACTERISTICS PERTINENT IN CHARGING

Important charging features of batteries are:

- The charging voltage is always higher than the discharging voltage at a given state of charge. The voltage difference, which depends on current density, represents a loss.

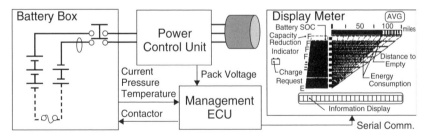

Figure 4.1 The Honda EV Plus carries a "management electronics unit" that supplies the driver a display of his distance of travel available with the state of charge of the battery, and other data.

- The life of a battery in terms of number of charge–discharge cycles generally varies with depth of discharge. Deep discharges shorten cycling life.

The different characteristics of candidate batteries for electric bicycle propulsion are summarized in Table 4.1. The terminal voltage of any battery type varies with the current flow, during both charge and discharge activity. The values shown in the table are nominal values. This voltage at different current levels also varies with temperature in a unique manner for each battery type. Lithium batteries had a limited life in charge/service at the time that this chart was created. Subsequently, the causes of their limited life have been discovered, and lifetimes of over 25,000 charge–discharge cycles have been demonstrated.

4.4   LEAD–ACID BATTERY CHARGING

A battery cell discharges when the negative plate can deliver electrons through a conductor to the positive plate (cathode). The lead–acid negative battery plate (anode), during discharge, absorbs sulfate ions from the electrolyte, forming lead sulfate and releasing hydrogen ions that drift to the positive plate to form lead sulfate and water. The battery is fully discharged when its plates are covered with lead sulfate.

TABLE 4.1   Battery Characteristics Pertinent to Recharging [a]

Battery Type	Nominal Cell Voltage (V)	Nominal Charge Voltage (V)	Deeply Discharged State (V)	Depth of Discharge (%)	Life Cycles	Storage Capacity Loss
Lead–acid	2	2.4	1.5	20	1500	0.27%/day [b]
				80	80	0.174%/day [c]
						0.035%/day [d]
NiCd	1.2	1.4	0.8	20	500	1.56%/day [e]
				20	16500 [h]	
				80	625 [h]	
NiMH	1.2	1.4	0.8	20	500	1.56%/day [e]
				80	6000 [h]	
Lithium	3.4 [g]	4.2	2.5–2.7	20	1000	0.33%/day [f]
				80		
Zinc–air	1.2	Not applicable	0 [i]	Na	200	Na

[a] J. S. Enochs and his colleagues reported achieving over 2000 charge–discharge cycles with 80% depth of discharge with lead–calcium–tin grids. (From J. S. Enochs, Nonantimonial Lead-Acid Batteries for Cycling Applications, Proceedings of the 19th Intersociety Energy Conversion Engineering Conference, ANS, 1984, pp. 850–856.)
[b] Standard grid.
[c] Low antimony grid.
[d] Calcium lead grid.
[e] Self-discharge is highest within first 24 h, 6% for NiCd and 9% NiMH.
[f] Includes 0.1%/day for self-protection circuits.
[g] At 50% of capacity, 3.0 V at 20% of capacity.
[h] Space application NiCd batteries.
[i] When zinc is consumed, there is no voltage.

A simple charger for lead–acid batteries is easy to design and build. It consists of a transformer that reduces the public power supplied voltage and a rectifier that converts the alternating current to direct current at the battery charge voltage. Low-cost chargers are available at stores that sell computers, automobile repair parts, and amateur radio equipment.

The options available for charging a lead–acid bicycle propulsion battery illustrate the range of available characteristics available in battery-charging technology and how their possible benefits can be evaluated. For example, the battery can be recharged with a transformer–rectifier that is plugged into an ac public outlet and designed to deliver to the battery all the current needed to maintain a voltage of 2.4 V per cell until the battery is fully charged and accepts no further current. At this point the cell voltage is maintained at 2.37 V per cell.

One common technique for recharging storage batteries is simply connecting the battery terminals to a dc voltage source that has a voltage that is greater than the battery voltage. This voltage difference will cause a charging current to flow through the battery and reverse the chemical reaction that occurred during discharge. The charging current decreases as the voltage difference between the charging voltage and the battery voltage decreases. Typically, the selected charging voltage is greater than the nominal battery voltage in order to cause a slight overcharge of the battery. The battery is deemed to be "charged" when the battery will accept no additional current. Most battery chargers have an ammeter, and the user is instructed to switch off the charger when the indicated current falls to zero. This constant-voltage charging technique is relatively safe since as the charging process progresses, the charging current decreases until it is just a trickle. Constant-voltage chargers are designed for overnight restoration of a discharged battery to a fully charged condition.

An alternative for quickly recharging a battery is a constant current charger that varies the voltage that is applied to the battery terminals in order to maintain a constant current flow. The charger automatically raises its voltage to keep the constant current flowing. The charger contains a controller that monitors the battery voltage and current flow. When the battery reaches full charge, the controller stops the constant current charging. Otherwise, overcharging would permanently damage the battery and might even boil of the battery electrochemicals. Bertness [1] uses a controlled current bypass circuit in a constant current source. The charging current is divided between the battery being charged and the bypass circuit. Both the battery voltage and its temperature are used in controlling the bypass current.

Other charge control methods are similar to those described for lithium and nickel-based batteries. For example "chopped" current waveforms can change the effective value of charging current. Control circuitry can also limit charging to a low "trickle" level if the battery had been completely discharged. Otherwise the relatively low battery resistance for this battery condition could overload the charger. As the battery gets charged, this resistance rises so that full charging current can be applied.

4.5   CHARGER DESIGN FOR LONG BATTERY LIFE

Another problem with battery charging is that the temperature of the battery typically rises during the recharging cycle. As the temperature of the battery increases, its chemical reactivity increases. The reactivity approximately doubles for every $10°C$ temperature rise in lead–acid batteries. Furthermore, as the temperature of the battery increases, its internal resistance decreases so the battery will accept a higher charging current at a given charging voltage. Bertness [1], uses the function

$$V = 14.32 - 0.024°C \qquad (4.1)$$

as the voltage to be applied to an automobile lead–acid battery. The term $°C$ is the battery temperature in degrees Celsius (equal to or greater than 0 for the voltage function). The decreasing voltage function is used to decrease an otherwise constant charging current with increasing battery temperature. Other types of lead–acid batteries can be charged with different linear functions by changing values of resistors within the Bertness charger. If it were not regulated, the increased current flow would generate additional heating in the battery, further reducing its internal resistance. This battery heating, followed by an increase in battery charging current, could result in a runaway condition that can damage the battery.

The direct current for the charger is obtained from a power conditioner. The power conditioner that rectifies the source current is connected directly to the public ac power supply. It then converts the power to a high-frequency (typically in the order of 25 kHz) pulsed current that goes through a step-down transformer into a rectifier that produces direct current for charging the battery. This approach decreases the size of the transformer because it transforms higher frequency power.

Discharging less than the full capacity of a battery during each use, plus sophisticated charge control, can make a lead–acid battery that is designed for bicycle propulsion have a lifetime of many years. For example, the battery carried on a bicycle needs to be a sealed type to avoid the damage that the leaking sulfuric acid electrolyte could cause if the bicycle were laid on its side or suddenly accelerated on bumpy roads or in maneuvers. Also, the bicycle might be ridden in cold weather. Therefore, the charger would need to recognize that the battery is cold and modify the profile of its voltage–current output appropriately. By using integrated circuit technology that is available today, a lightweight charger can be programmed to make a battery have its longest possible life.

4.6   SMART CHARGERS FOR NEW NICKEL–CADMIUM, NICKEL–METAL HYDRIDE, AND LITHIUM BATTERIES

There is a growing desire by users to charge batteries quickly. However, batteries do not react rapidly to either charging or discharging. The faster batteries are charged, the less total energy they will accept before reaching voltage or temperature limits. Surpassing these limits either causes damage or reduces battery life.

There are three regimes of time for battery charging: slow, quick, and rapid. The slow rate is generally at a current rate of $C/10$ or less, where C is the battery's rating in ampere-hours. The quick rate is generally at a current rate of around $C/3$ and the rapid rate is $C/1.5$ or higher [2]. These rates correspond to time spans of 10 h or more for "slow," around 3 h for "quick," and 1 h or less for "rapid." There is also a maintenance mode for which the battery is connected to the charger when it is not being used for a long period.

Battery-charging systems have ranged from a simple transformer–rectifier type to complex systems that monitor and control the charging function. To reliably and efficiently charge NiCd, NiMH, and lithium batteries at high rates requires careful control of the charging operation to avoid damage to the cells, particularly under extreme ambient temperature conditions.

4.6.1 Problems to Overcome

A characteristic of the nickel–cadmium and nickel–metal hydride battery cells was their fall-off in accepting charge during the charging process. Not all of the current supplied to the battery is recoverable or utilized in the chemical reactions by which the battery is charged. The percentage of incremental input current, or charge, that is recoverable at any given point in the charge cycle is referred to as the charge acceptance of the battery. The cumulative charge accepted by the battery determines the battery state of charge in terms of the percentage of full charge [3].

To fully recharge the cells, as much as 160 percent of their rated energy capacity needs to be replaced [4]. This extra charging energy is dissipated as heat. The added heat can raise the cell temperature to the point of cell damage because of the relatively high thermal resistance of batteries. Generally, the input charging current supplied to a battery must be limited in order to prevent an overtemperature condition. It has been shown that charging a NiMH battery at a safe $C/10$ rate only 50 percent of the charge was accepted within 10 h [4]. The battery required 16-h to reach full charge at the $C/10$ rate. Higher current charging is required to shorten the charge time, but higher currents also raise the battery temperature. For this reason, the battery temperature is sometimes measured and used to trigger a reduction in the charging current. Also as the battery temperature rises, the charge acceptance is degraded, often resulting in an incapability of the battery to reach its fully charged state.

The most critical factors in determining the maximum allowable charge current that can be safely delivered to these batteries are temperature and state of charge. At low temperatures the oxygen recombination rate is significantly reduced. This limits the allowable overcharge current that may be applied without venting the cells if they are fully charged. At high temperatures the heat released by the oxygen recombination reaction may cause excessive cell temperature leading to premature failure of the plate separator material and possibly a subsequent short circuit.

A nickel–metal hydride battery, in particular, has a temperature rise gradient that varies greatly with charging current and the already charged capacity. Furthermore, the nickel–metal hydride battery is less resistant to overcharge than is the nickel–cadmium battery. If overcharged, the nickel–metal hydride battery's life is shortened [5].

Figure 4.2 shows an example region of acceptable and unacceptable rates of charge as a function of battery temperature. If overcharging occurs, the current will cause generation of oxygen gas with only a relatively small amount of charge actually being stored in the cell. If the charge rate is too high, the rate of oxygen recombination that occurs within the cell may be insufficient to prevent excessive internal pressure and cell venting, which drastically reduces the useful life of the cell. If the battery is fully discharged, minimal oxygen generation will occur until the battery nears the fully charged condition. And when the battery is nearly fully charged, it can quickly enter the overcharge condition and begin oxygen generation. The difficulty lies in accurate determination of the preceding state of charge to avoid damage to the battery.

Repetitive shallow discharging of the cells causes another problem with NiCd cells. Repetitive shallow discharge progressively reduces the capacity of the cell. Figure 4.3 illustrates this.

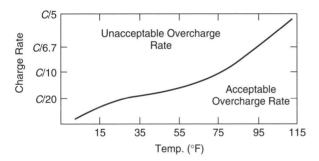

Figure 4.2   Example regions for nickel–cadmium and nickel–metal hydride cells of acceptable overcharging rate in terms of battery capacity as a function of battery temperature [7].

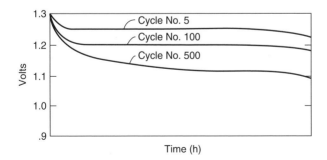

Figure 4.3   Example that shows reduced NiCd cell capacity after repetitive cycles of shallow discharging [7].

Lithium-based batteries can be dangerous if overcharged, in contrast to either lead–acid or nickel-based batteries. An overcharged lead–acid battery will electrolyze easily replaced water, and nickel–cadmium or metal hydride batteries have voltages that stop rising at full charge. However, the voltage of a lithium–polymer battery cell continues to rise even while being overcharged, and lithium ion cells carry a risk of generating excess gas due to overcharge or overdischarge. Lithium metal is explosive in water and will, in varying degrees, react with the moisture in the atmosphere. Lithium-containing batteries have been known to explode or catch fire, although more recent safety designs have reduced the chances of this occurrence. In the lithium–polymer battery cell, wherein lithium ions are contained in a solid polymer, there is no liquid to vaporize. Overcharging depletes the lithium ions off of the plate, thus breaking its electrical connection.

The life cycle of the cell is decreased even if a catastrophe does not happen. The avoidance of overcharge voltage and overcharge current during charging of a lithium-based cell is therefore an important objective in the use of such batteries.

Another problem is the equality between cells in a multicell battery. There can be many causes for differences between cells and, consequently, their state of charge. In particular, larger cells being developed for electric vehicle applications are not as consistent from cell to cell as are the mass-produced commodity cell in laptop computers [6]. Some cells can be overcharged while others are undercharged if the battery is charged without regard to the condition of each cell. To avoid this dangerous condition, cells can be individually monitored so that no cell is overcharged. The critical dependence on voltage for the lithium cell makes series-connected cells more difficult to handle because the voltage on each cell must be monitored during charge and discharge. Action is required when the highest voltage cell reaches maximum or when the lowest voltage cell reaches minimum. Reducing the charging voltage on one lithium cell by just 100 mV reduces its capacity by 10 percent or more [7].

4.6.2 Charging Solutions for Nickel–Cadmium and Nickel–Metal Hydride Batteries

A low-cost way of charging batteries is to use a simple *trickle charger*, which restores a discharged battery to full charge at a rate of around $C/10$ [2]. This charger, which requires at least a 10-h charge period, can be used to recharge electric bicycle batteries if the user is careful. This simple charger has to be turned on and off by the user. However, if the user is not careful, the battery can get over-charged. One means of control is to use a timer that turns off the trickle charger after the battery capacity consumed during motor-driven travel has been replaced. A simple criterion would be to assume the battery is used at some average fraction of the battery's C capacity per hour during travel. The user would then set the timer to turn off the charger after a time equal to 10-to-16 times the motor-driven travel time. This technique might be especially applicable for commuting to work. The energy consumed in a half-hour motor-on commute travel could be nearly restored by trickle charging for 8 h. The commuter could plug in the trickle charger when she arrives at work and unplug it at the end of the workday.

Battery charging is often characterized by three phases: (1) an initial charging phase during which the charge acceptance is relatively high, (2) an intermediate charging phase when the charge acceptance decreases, and (3) a final charging phase during which the charge acceptance approaches 0 percent and the battery has reached its full charge state. One type of conventional battery charger circuit decreases the level of input current supplied to a battery at the transition between such phases, based on battery voltage. However, even with careful selection of the input current, this technique can only provide fast charging consistent with safe battery temperature during a portion of the charging phase. This is because the charge acceptance varies continuously during battery charging. Thus, a constant level of input current, no matter how carefully selected, will not be optimum throughout an entire charging operation [3].

Detecting the completion of charge is typically based on laboratory measurement of the battery charge and temperature characteristics. These values are then used in battery-charging algorithms stored within the charger control electronics. Algorithms vary among manufacturers from simple to complex. With this method, however, if the number of battery cells in a battery pack increases, a battery cell positioned on the end portion of the battery pack tends to be cooler than the cell in the center. The result is that the completion of charge cannot be detected based on one battery temperature alone. Furthermore, if the temperatures of the cells in the battery pack become increasingly disparate, the life of a higher temperature cell is shortened. Also the higher temperature cell is more difficult to charge. Due to this, the capacities become unbalanced among the cells, and it may happen, for example, that there is a cell charged 100 percent and another to 90 percent. Hence, if the capacity of the battery pack is used up to 90 percent, some cells will have a residual capacity of 10 percent and the others 0 percent. The cells having 0 percent residual capacity can be reverse charged by those of 10 percent residual capacity, and the battery life would be shortened.

Charging by Means of Voltage Detection   Various methods have been proposed to detect the fully charged condition of a battery. One is to sample battery voltage. When the peak voltage appearing at the charge termination period is detected, the battery is assumed to have reached a fully charged condition. This procedure, called *peak voltage detection method*, is shown in Figure 4.4. Another method is to detect battery temperature and compute a rate of temperature rise. This is called *temperature rise gradient*. When the temperature rise gradient has exceeded a predetermined value, the battery is considered to be fully charged. This method is referred to as *dT/dt detection method* (Fig. 4.5) [5].

The peak voltage detection method, a potentially inexpensive alternative, is not suitable for batteries that exhibit battery charge characteristic with no clear peak voltage [5], such as the lithium battery. Charging NiMH and NiCd batteries with a voltage source set too high would result in large currents that would overcharge, damage, and perhaps destroy the battery. In addition, NiMH and NiCd cells have negative voltage temperature coefficients. In overcharge the battery heats up and its voltage decreases, making the problem even less manageable. However, lead–acid car batteries, vented NiCd, and similar batteries

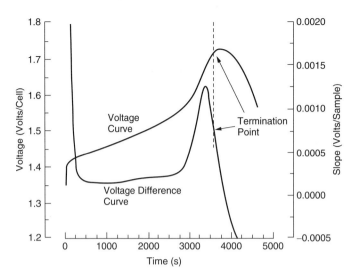

Figure 4.4 Illustration of battery voltage change detection used for charge termination. The "voltage difference" curve is the first derivative of the voltage curve. There is a slight time delay in the derivative curve. (After George E Sage, Pulse-Charge Battery Charger, U.S. Patent 5,633,574, May 27, 1997.)

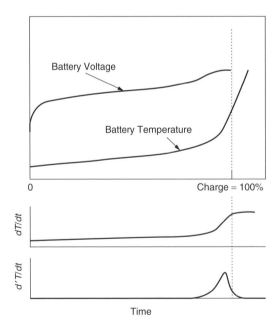

Figure 4.5 Illustration of battery voltage and temperature rise during charge, the temperature change with time (dT/dt) and the dT/dt change. The point of 100 percent charge is shown [5].

that are used in industrial applications, because of their differing characteristics, are quite amenable to voltage sensing [8].

Control of the duty factor and frequency of charging pulses, in effect changing the charging energy, can be achieved in three stages of charge: (1) fast charge,

(2) taper charge, and (3) trickle charge, can be used to solve the problems associated with peak voltage sensing [6]. The fast charge stage brings the battery voltage up to a precise predetermined value. It then turns off and remains off until the battery voltage drops to another precise predetermined value at which point it turns on again. The process continues. Each time as the battery accumulates charge the duty factor of the pulsing naturally reduces.

Charging by Means of Temperature Control   Although the dT/dt detection method is claimed to be superior to the peak voltage method, it alone may fail to detect the fully charged condition of the battery [5]. In the dT/dt detection method, the temperature rise gradient is compared with fixed critical value. As such, detection of the fully charged condition of the battery is made based, among other things, only on the temperature rise gradient. Other factors, such as the kind of battery to be charged, the condition of the battery, battery temperature at the time when charging starts, charge current, or ambient temperature, are not considered for determining the fully charged condition. Those unconsidered factors may increase the battery temperature rise gradient more than the preselected value, despite the fact that the battery has not yet reached the fully charged condition. In such a case, charging is stopped before the battery is fully charged, so the battery is undercharged. On the other hand, the battery temperature rise gradient may not increase more than the fixed critical value, despite the fact that the battery has reached its fully charged condition. In this case, the battery is overcharged because charging will not stop even if the battery is fully charged. Overcharging can cause electrolysis of the battery's electrolyte. The resulting gas pressure can force electrolyte to leak out from the battery. This shortens a cycle lifetime of the battery.

These problems can be overcome by using more of the battery parameters, some of which can be obtained from laboratory measurements and vendor data. These parameters are stored in the battery control electronics and in the algorithms used by the charger's electronics. Equations are used that include integrals of the state of charge, rated charge capacity of the battery cells, actual measured cell capacity, temperature characteristics recorded during laboratory tests, empirically determined temperature variables, plus cell temperature, current, and voltages during the charge operation. The charging algorithm [3] has these six states:

1. *An idle state.* This occurs either when the charger is powered, in response to a "start" command, or can potentially occur at other times during the charge cycle in response to battery condition measurements indicating that the battery is unsuitable for further charging.

2. *A precharge state.* This occurs if the battery is to be within safe limits for receiving a high charging current.

3. *A high current charging state.* This is the initial charging phase when the battery's charge acceptance is relatively high. This state ends when the cumulative input charge to the battery has reached approximately 90 percent of its rated storage capacity.

4. *A high-temperature state.* This occurs during the first portion of the intermediate charging phase, when the battery temperature is maintained at approximately 40°C.

5. Thereafter, *a low-temperature state* is entered. During the second portion of the intermediate charging phase, the battery temperature is maintained at approximately 33.5°C.

6. Finally, the battery enters *a maintenance state* where it is maintained in a full-charge state.

Ambient temperature, as well as cell temperature can be used to control charging. The difference between ambient temperature and battery temperature is used in a charging algorithm to select between a fast charge mode and a battery conditioning mode [9]. During the selection the battery temperature is adjusted relative to ambient temperature. These data are used in adjusting the charging current.

A resistor and associated temperature sensor within the battery charger for monitoring and controlling battery temperature is sometimes used [10]. This concept eliminates a sensing wire. The resistor temperature is calibrated prior to charger use with actual battery temperature for various values of charging current. However, it may be difficult to implement this technique for high current charging.

A method of controlling the temperature rise of the battery so that it stays within certain polynomial functions based upon battery types can be used to rapidly charge a battery [11]. The temperature functions can be linear, upwardly rounded, downwardly rounded during temperature increases, or double curved increase functions. With this technique one can obtain better and safer battery charging. A battery can be charged within 10 to 20 min using this method.

Charging by Means of Charge and Discharge Pulses It is possible to improve charging of nickel–cadmium cells with cell discharge pulses during the charge process [12]. The technique is suitable for recharging primary zinc–manganese dioxide ($ZnMnO_2$) alkaline cells, as well as rechargeable alkaline manganese cells, nickel–cadmium cells, and conventional standard and heavy-duty "dry cells" to the extent that they are rechargeable.

The technique uses dc charging pulses with durations of 1 to 8 ms. They are derived from normal ac utility power. The current pulses have an average value that is from about 9 to about 20 percent of the initial current capacity of the cell and do not exceed 40 percent of the manufacturer's rated values.

The charger uses a reference voltage for each battery type. The reference voltage is equal to or less than 1.7 V for an alkaline manganese cell. When the process is used with a lead–acid battery, the reference voltage limit is 2.45 V per cell. When the process is used with a nickel–cadmium cell, the reference voltage limit is 1.42 V per cell, which in this case represents the transition voltage between current limited and constant current charging. In general, the reference voltage must not exceed the long-term, safe float voltage of any cell.

Pulses of direct current are periodically withdrawn from the cell. The duration of these pulses range from 5 to about 35 percent of the duration of the dc charging pulses. They also have a current value during discharge of 10 to about 25 percent of the average current value available from the charging pulses. In general, the discharge pulses represent a loss of from about 3 to about 8 percent of the available charger energy. In addition, this charger periodically terminates the charging process and provides continuous discharge of current from the cell for a relatively long period of time. The duration of the "continuous" discharge pulse will be at least 1.0 s. The current withdrawn during the long-pulse discharge is generally from about 10 to 25 percent of the average current available from the charging pulses.

These are example times and ratios for charging and discharging. Specific battery chemistries vary the times and ratios. By comparison, with lead–acid batteries, the duration of the long pulses is from about 15 to about 30 s, and the time between long pulses is from about 1.5 to about 5.0 min. The technique is especially applicable to nickel–cadmium batteries that suffer from "memory effects."

4.6.3   Charging Solutions for Lithium Batteries

Two-Step Charging   The operation of charging is first to supply a charging current until a certain cell voltage is reached. Next, the cell voltage is held constant at that predetermined value until the charging current decreases to a particular value, about $\frac{1}{10}$ to $\frac{1}{20}$ of the constant current charging value. Figure 4.6 shows the changes in cell parameters during cell charging. I_0 is the constant current value and V_0 is the manufacturers cutoff threshold voltage. The voltage value accuracy is typically required to be 1 to 2 percent [13]. The chemistry of some common cells can cause this charging voltage to be 4.1 V $\pm$ 50 mV and 4.2 V $\pm$ 50 mV [14]. Six- or nine-cell batteries for 24- or 36-V electric bicycle systems, respectively, will necessitate a battery charger capable of supplying a charging voltage by as much as 6 or 9 times this possible voltage spread of

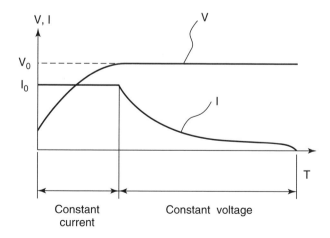

Figure 4.6   Constant current, constant voltage charging of lithium cell.

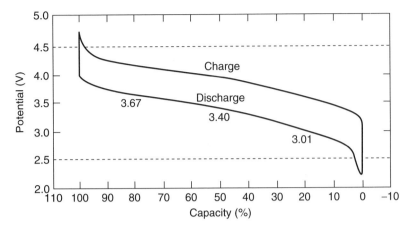

Figure 4.7 Cell potential vs. state of charge. (From Lauren V. Merritt et al., Lithium Polymer Battery Charger Methods and Apparatus, U.S. Patent 5,773,959, June 30, 1998.)

100 mV. That amounts to about a 0.5- to 1.0-V difference. Failures to adjust this difference could cause battery damage.

During discharge a lower cutoff potential of 2.5 V or above avoids damage to a cell. Figure 4.7 shows the details of a cell potential as a function of state of charge during charging. During charging this cell's nominal 4.2-V potential is obtained when about 80 percent of its capacity has been restored, and the remaining capacity is recovered during the subsequent constant-voltage charging. Test data of various lithium ion cell technologies shows that the 4.2 V is achieved at between 40 and 70 percent of capacity. This variation is a function of the constant current value during charging [15]. Doubling the current during the constant current-charging phase reduces the charging time by less than 30 percent [16].

Three-Step Charging The conventional battery chargers that use the two-step charging process may cause a problem if the lithium battery is too deeply discharged. The typical battery charger contains charge control components that vary controlled impedance that is in series with the battery to control the charging of the battery. The controlled impedance may dissipate an excessive amount of power during the early stages of constant current charging, depending on the discharge state of the battery. A discharged battery has a relatively low terminal voltage compared to a fully charged battery. Therefore, during the early stages of constant current charging a higher voltage is impressed across the controlled impedance, resulting in increased dissipation. The increased dissipation leads to reduced reliability and possible failure of the controlled impedance.

This problem is avoided by controlling the charging current during the initial charging operation so that the power dissipated in the charge circuit is

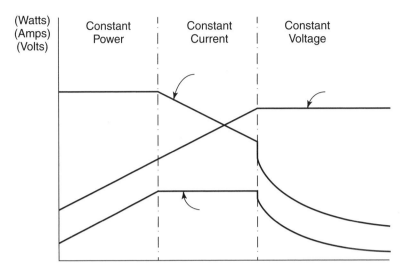

Figure 4.8   Constant power, constant current, constant voltage charging of lithium cell [16].

held constant [16]. Thereafter the charging voltage is held constant as shown in Figure 4.8. The power curve shown is for the power dissipated within the charger and not within the battery. The power dissipated and absorbed by the battery is the product of the battery currents and voltages shown for the three stages of charging.

Additional problems arise if the constant current is held at values greater than the cells $1C$ rate. One is that metallic lithium will be plated onto the electrode rather than being adsorbed in the electrode. This can permanently reduce the capacity of the cell [17]. In an attempt to shorten the charging time, some chargers are being designed to increase the charging voltage to force a higher charging current during the charge current's trickle-down phase. This results in dissolution of the electrolyte, the plating of metallic lithium, and a consequent shortening of the battery life. The lifetime of the batteries is reduced, generally to about 300 cycles, and the charging time is only reduced to 3 to 4 h [17].

Charging with Pulses and Rest Periods   The negative effects discussed can be reduced by using sequences of pulses that are separated by rest periods, and both charge and discharge the cells [17]. The overall effect is to decrease the charging time. Four sequences can be used: The first is a charging phase that uses three to five charge pulses within about 25 ms, separated by rest periods. The second is a "removal" phase of alternating charge and discharge pulses. The third is another "removal" phase that uses large magnitude discharge pulses followed by rest periods. The fourth is a "measurement" phase where cell voltage, impedance, and temperature are determined. The measured values are then used to alter the parameters and the application of sequences used for subsequent charge cycling.

This method reduces the buildup of metallic resistive lithium on the positive electrode, minimizes dendritic formation that causes short circuits, and minimizes the electrolyte decomposition. These techniques are applicable to lead–acid and nickel–hydride batteries as well as lithium batteries.

4.7 SMART BATTERIES FOR SMART CHARGERS

The stringent charging requirements for lithium batteries and also the nickel–metal hydride batteries created a need for what are termed *smart batteries*. A smart battery is a battery or a cell that contains electronics that are capable of reporting the present and optionally the historical characteristics of that battery or cell. There is no standard list of characteristics to be reported or methods of implementation. Consequently, smart battery designs vary among their manufacturers and the battery applications [7, 18]. Since lithium cell chemistries can vary, one would desire to have the manufacturer's threshold charge cutoff voltage for the cell to be reported to the battery charger. Another drawback of present battery-charging techniques is that no accurate indication is provided on how long it will take to fully charge the battery from its present state of charge. For example, a half-discharged battery could be restored to full capacity more quickly than can a fully discharged battery.

For propulsion applications a "gas gauge" type of battery state-of-charge reporting is very useful. Displaying the accumulated number of charge–discharge cycles would enable the bicycle owner to estimate the remaining life of the battery. The owner of a battery composed of expensive cells needs to know when one cell needs to be replaced before it causes a complete battery failure. Today's battery chargers are not capable of being adapted to meet changing charging needs as battery chemistries and cell designs change over time.

A smart battery can predict whether or not a battery can deliver a requested amount of additional power, based on the battery's specific capacity, self-discharge, and discharge characteristics [19]. Figure 4.9 shows a block diagram of the circuitry placed in a battery. This block diagram can be generally characteristic and applicable to many smart batteries but without the display. Some smart batteries will contain only a nonvolatile read-only memory. A nonvolatile memory stores the battery-specific characteristics, which are functions of the environmental conditions of the battery and the battery current. In response to a request, the smart battery determines whether or not the requested additional power can be provided, based on the present battery capacity, the present discharge rate of the battery, the environmental conditions of the battery, and the battery characteristics.

It is possible to make the smart battery even control its own charging [20]. Furthermore, it constantly monitors the battery voltage, temperature, current

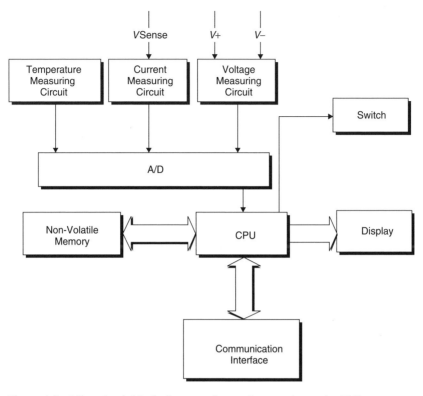

Figure 4.9   Microcircuit block diagram of smart battery electronics [21].

charge–discharge rate, and remaining capacity of the battery. This smart battery system is capable of being charged from a simple voltage supply. All aspects of charge–discharge rate control and monitoring are accomplished by the smart battery system, rather than in an external charging circuit. This eliminates the need for external systems that include such circuitry and control elements.

A smart battery design can maintain information for indicating when the smart battery requires maintenance [22]. A battery maintenance and testing system can read this need for maintenance from the smart battery and then proceed to maintain the smart battery. Conditions that indicate that the battery is defective or has exceeded its useful life are also recorded by the smart battery and shown on a display or delivered to other devices over a communication bus.

The smart battery can be made to contain a magnetic-stripe reader, an integrated circuit (IC) card reader, and a Personal Computer Memory Card International Association (PCMCIA) card reader [23]. Finally, a microcontroller interfaces a cellular telephone with the reading devices. A universal remote control device, such as those used for controlling the normal television or home entertainment system recorder, can be connected to the microcontroller. It enables the use of the keyboard of the cellular telephone to control operation of the universal remote control device. Furthermore, the microcontroller will be capable of transmitting information toward a remote central processor through the cellular telephone.

4.8 SELF-DISCHARGE RATE OF NICKEL AND LITHIUM CELLS

It is often important to provide accurate information on the remaining capacity of the battery. Some batteries have a "fuel gauge" that indicates the charge level of the battery. One example is a rechargeable battery that contains an internal fuel-gage that measures its discharge current and estimates its self-discharge to predict the remaining capacity of the battery [24]. This corresponds to a fuel gauge that reports the content of a "fuel tank."

Self-discharge refers to a loss of battery capacity that occurs even when the battery delivers no discharge current to a load. In Dunstan [25] a fully charged NiMH battery is estimated to self-discharge at a rate of approximately 6 percent in the first 6 h, and a NiCd battery is estimated to self-discharge at a rate of 3 percent in the first 6 h. Both batteries are estimated to self-discharge at rates of about 1.5 percent in the second 6-h period, about 0.78 percent in the third and fourth 6-h periods, and approximately 0.39 percent in each subsequent 6-h period. For partially discharged batteries, self-discharge is estimated to be about 0.39 percent for each 6-h period. It is inferred from this data that after the first day NiMH cells will have lost about 9 percent of their charge, and NiCd cells will have lost 6 percent of their charge. After 60 days each battery will be fully discharged.

The self-discharge rate for nickel–hydrogen cells has been reported to be 8 percent per day for batteries that start with 90 percent state of charge. Similarly, nickel–cadmium cells had a self-discharge rate of 2 percent per day. These rates apply for a temperature of 10°C. As the cells discharge the rate of self-discharge slows.

Zimmerman [26] showed self-discharge rates for various capacity lithium ion cells. The average rate applied to cells starting from cell voltages in the range of 3.8 to 4.2 V. These are the voltages of the cells that had been standing on the shelf for 185 days. Data for the various batteries discussed are shown in Table 4.1.

We can generalize that the self-discharge rate of lithium cells follows the function:

$$S = C^{0.5}/60,000 \tag{4.2}$$

where S is the average self-discharge, expressed as a fraction of the cells capacity per day, C is the capacity of the cell, in ampere-hours. The constant of 60,000 was determined for a range of cell capacities from 1.24 to 40 Ah. It is concluded that during a 60-day period the fully charged lithium cell will loose 0.2 percent of its stored energy.

4.8.1   Applicability of Commercial Fuel Gauges to Electric Bicycles

A commercial fuel gauge approach for predicting battery capacity does not account for the dependence of self-discharge or battery capacity on environmental conditions, such as battery temperature. In certain situations, methods [25] for estimating self-discharge can produce unacceptable errors in the fuel gauge that indicates remaining capacity. In high-temperature conditions the self-discharge can be much greater than the estimated values. For example, one scenario could include a video camcorder and battery being stored in a car during a series of hot, sunny days. This would cause the battery's fuel gauge to erroneously indicate more than enough battery charge for filming a 20-min wedding ceremony. The scenario concludes with an irate father whose camcorder shuts off after 10 min of the ceremony, and he had no spare battery.

The above-described fuel gauges are used to indicate a battery's capacity and display an alarm when a battery's energy content approaches exhaustion. However, for use on electric bicycles, these presently available fuel gauges have the following difficulties: First, the presented discharge rate does not adequately reflect the dynamic discharge rates an electric bicycle produces. Second, the battery capacity provided by the fuel gauge is inaccurate because of the dependence of self-discharge on environmental factors, such as temperature, humidity, and air pressure, which were not considered. Third, the alarm value is fixed. This removes flexibility from the power management system for adjusting the alarm value to the varying power conditions in the system.

4.9   RECOVERABLE ENERGY

We often hear the question "When going down hill does your system recharge your batteries?" Or "Does your bike have regenerative charging?" To answer these questions a computer analytical model was used to predict when there was excess energy available from the motor. Several viewpoints were taken to find the answer. The first was to assume specific hill grades, head-wind conditions, and bicycle conditions. The power that could be available from the motor after accounting for air resistance and rolling resistance was determined. The second viewpoint was to determine analytically, in the same model, the energy recoverable for an actual travel route. A 200-mile (322-km) Seattle-to-Portland route, which local bicyclists ride in an annual event, was used in the analysis. One can also use the section in Chapter 2 on acceleration to help answer questions about regenerative braking. Each viewpoint leads to answers that can help decide how important it is to recharge the batteries with the kinetic and potential energy of a bicycle.

Some of the results obtained for the first viewpoint, where particular grades, head-wind conditions, and bicycle conditions that were assumed are shown in Table 4.2. This table shows three items of interest: (1) The rate of regenerative energy per kilometer of distance obtainable from a motor with 100 percent recharging efficiency, (2) the bicycle downhill speed, and (3) the bicycle speed without regenerative charging. In reality, 100 percent recharging efficiency is not possible.

TABLE 4.2 Example Computations of Electric Bicycle Regenerative Charging

Grade (%)	Head Wind (km/h)	Maximum Speed[a] (km/h)	Maximum Power (W) Recoverable	Speed at Max. Regenerative Power (km/h)	Recoverable Energy[b] per km (Wh)
−3	0	37	122	21	5.8
−3	10	27	72	15	4.8
−6	0	55	407	32	12.7
−6	10	46	294	26	11.3
−9	0	68	789	40	19.7
−9	10	59	611	34	18
−12	0	80	1246	47	26.5
−12	10	71	1003	40	25.1

[a]Maximum speed attainable without regenerative braking.
[b]100% efficiency assumed after attaining coasting speed.

A realistic number is between 40 and 60 percent efficiency. The table values were calculated for an electric bicycle and rider total weight of 122 kg (269 lb), coefficient of drag is 1, rolling coefficient is 0.0058, frontal area 0.5 m^2, and air density for an elevation of 15 m. Energy recovery was assumed to be 100 percent efficient.

As an example, if we assume 50 percent recharging efficiency, we will get about 6.3 Wh/km (10.1 Wh/mile) traveled at a steady-state speed on a downhill grade of 6 percent. This amounts to an energy recovery of 2.5 percent for a 250-Wh battery. The steady-state coasting speed with this regenerative charging, under the conditions described in the caption for Table 4.2, is about 32 km/h. Without regenerative charging the speed would be 55 km/h. If one had many 1-km coasting distances, battery recharging would occur.

The distribution of road grade and length needs to be considered in assessing the benefit of recovering energy. As a further example again use the route in the annual Seattle-to-Portland bicycle run [27]. In this approximately 200-mile (322-km) distance we found that 96 percent of the route had road grades of less than 6 percent and 91 percent of the route had grades of less than 4 percent. By accounting for length of the grade, we obtained the amount of recoverable energy for each downhill segment. Results are shown in Figure 4.10. For any particular grade the data points are scattered due to varying length of grade and various wind and efficiency conditions are assumed. Little beneficial recoverable energy is shown for downhill grades less than 3 percent.

A generalized consideration might follow the approach: Assume it takes on the average 15 Wh/km (24 Wh/mile) to travel a certain route and that the average downhill grade is 4 percent. Furthermore assume, using a route similar to the Seattle-to-Portland route, that the downhill grades for charging are 5 percent of your route. Assuming 4 Wh/km of regenerative energy for the average 4 percent grade. The percent of recoverable energy for this example is

$$5\% \times (4 \text{ Wh/km}/15 \text{ Wh/km}) = 1.3\%$$

A more exact way would be to model each small increment of the route.

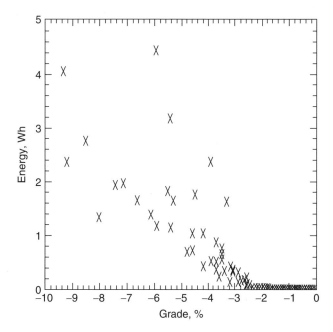

Figure 4.10   Recoverable energy from downhill coasting on the Seattle-to-Portland Bicycle Annual Classic. Note that for a given grade the length of the recovery zone affects the energy recovery. Vehicle and rider combined weight is 142 kg and coefficient of drag is 1 [27].

These example results show that the benefit of battery charging by energy recovery would be difficult to justify on the basis of the added cost, complexity, and weight. For this consideration, an inverter or converter is likely to be necessary because any motor when used as a generator will not supply the same voltage to its supply as when it runs as a motor. The difference is caused by the losses in the circuit and the torque losses in the motor/generator. In addition the charging voltage on the battery will have to be higher than its load voltage. The inverter or converter would also need good regulation so as not to put an overvoltage on the battery for the varying downhill travel speeds.

A perspective you might want to consider when addressing the subject of regenerative battery charging is to compare the fast moving heavy electric automobile to the slow moving and light electric-powered bicycle. The automobile has much more kinetic energy than the bicycle. The equation for the energy of a moving object is

$$E = \tfrac{1}{2}mv^2 \qquad (4.3)$$

where m is the mass and v the velocity. The car will typically be traveling 4 to 5 times the speed of the bicycle, and the mass of the car will typically be 10 times that of the bicycle. You can see that the kinetic energy of the car is $10 \times 4^2 = 160$ times that of a bicycle. The rate of energy use by an electric car is about 10 times that for an electric-powered bicycle. So in effect there is about a 16 to 1 improvement in regenerative battery charging for an automobile.

4.10   SOLAR PANEL BATTERY CHARGERS

Sunlight is often proposed as a nonpolluting source of energy for propelling vehicles. These proposals are accompanied by action-motivating text, and pertinent data for evaluation are not often provided. An illustrative example is a recent cross-Australia race of solar-powered electric cars. A General Motors car that had solar panels on its roof won the race. It arrived at the end point of the race 2 days before the nearest competitor arrived!

Not mentioned or shown in the publicity was the composition of the team and vehicles that made winning the race possible. The solar-powered car was accompanied on its route by two vans and an observer-carrying helicopter. One van carried the director who was supported by a computer that tracked the solar panel power output, the vehicle battery's energy content, and the propulsion power being consumed. He was also supplied wind-velocity data and the grades of hills that lay ahead. The observer in the helicopter would survey the geography of the route ahead of the solar-powered car, searching for areas where direct sunlight would not be available. The second van carried food for the race crew and tents for sleeping at night.

Early in one afternoon the director learned that the next morning's travel required the solar-powered racecar to cross a high mountain range. It first had to travel through valleys and forested areas where direct solar radiation was not continuously available. Only by starting with fully charged batteries could the car cross the range in one day. The director then ordered the helicopter pilot to locate, within a designated area, a camping spot where the solar car could be parked on a west-sloping slope. There the rays of the sun would arrive as close as possible to 90 degrees with respect to the solar panel during most of the afternoon. This decision enabled the racecar to cross the mountain range the next morning.

Bicycles in which battery power supplements pedal power have been supplied with solar panels. A solar panel can be oriented to collect the maximum sunlight when the bicycle is parked. A solar panel that tracks the sun could be installed on a bicycle. However, an analysis showed that even at moderate speeds there are many orientations of this panel at which the power required to overcome the wind force on the panel exceeds the power produced by the fully illuminated solar panel.

Our analysis did show that there are places where solar power can be a practical source of energy for propelling bicycles. For example, the African nation of Zambia has a need for quickly establishing agriculture and other colleges where skills can be developed for advancing the quality of living in the nation. Time and money were not available for quickly building dormitories or automobile roads on which students could commute to colleges. The student candidates do not have money for buying motorcycles that could travel on trails. Also the nation does not have the resources required for building a network of roads and pipelines that deliver petroleum fuel.

An economically practical student transportation system consists of bicycle trails, plus bicycles powered with zinc–air fuel cells. A student living 30 miles from the college could get to her class from home in 1.5 h. At the college she would stop by a "filling station" where she would drain the zinc–dioxide from the fuel cell, add fresh electrolyte into the cell, and pour zinc powder into the cell's hopper. An electrolysis unit at the filling station would restore the electrolyte by recovering the zinc.

Sun-following solar cell panels would generate enough power to enable the electrolysis unit to produce the daily fuel cell needs, plus a stored supply for cloudy days.

REFERENCES

1. Kevin I. Bertness, Battery Charge Control Device, U.S. Patent 6,696,819, February 24, 2004.
2. Bill Bently, Basics of Rechargeable Battery Management, *PCIM Magazine*, October 1995 pp. 66–70.
3. Peter R. Holloway and Robert A. Mammano, Battery Charger Circuit Including Battery Temperature Control, U.S. Patent 5,504,416, April 2, 1996.
4. Robert A. Mammano, *Charging the New Batteries—IC Controllers Track New Technologies*, IEEE 0-7803-2459-5/95, 1995, pp. 171–176.
5. Nobuhiro Takano and Shigeru Moriyama, Battery Charger and Method of Detecting a Fully Charged Condition of a Secondary Battery, U.S. Patent 6,335,612, January 1, 2002.
6. Jean-Pierre Vandelac, Compact Fast Battery Charger, U.S. Patent 6,301,132, October 9, 2001.
7. Phillip Miller and Ronald D. Becker, Fast Battery Charger, U.S. Patent 5,363,031, November 8, 1994.
8. Randall Wang, Auto-controller for Battery Charger Using Thermo-control and Current Balance Technology, U.S. Patent 6,404,169, June 11, 2002.
9. Kazuyuki Sakakibara, Battery Charger and Battery Charging Method, U.S. Patent 6,476,584, November 5, 2002.
10. Robert S. Feldstein, Alkaline Battery Charger and Method of Operating Same, U.S. Patent 5,523,667, June 4, 1996.
11. C. Liao, LT1510 High Efficiency Lithium-Ion Battery Charger, Design Note 111, Linear Technology, Milpitas, CA.
12. F. Hoffart, Li-Ion Battery Charger Adapts to Different Chemistries, *Electronic Design News*, September 2, 1999, p. 146.
13. Thomas T. Sack, J. Croydon Tice, and Ran Reynolds, Segmented Battery Charger for High Energy 28 V Lithium Ion Battery, *IEEE AES Magazine*, September 2001, pp. 15–18.
14. F. Goodenough, Battery-Management ICs Meet Diverse Needs, *Electronic Design*, August 19, 1996, pp. 79–96.
15. V. L. Teofilo, L. V. Merritt, and R. P. Hollandsworth, Advanced Lithium Ion Battery Charger, *IEEE AES Systems Magazine*, November 1997, pp. 30–35.
16. Kevin Dotzler and Keisaku Hayashi, Lithium-Ion Battery Charger Power Limitation Method, U.S. Patent 6,664,765, December 16, 2003.
17. Yury, Mikhail L. Podrazhansky, and Richard C. Cope, Battery Charger with Enhanced Charging and Charge Measurement Processes, U.S. Patent 6,232,750, May 15, 2001.
18. A. W. Swager, Smart Battery Technology: Power Management's Missing Link, *Electronic Design News*, March 2,1995, pp. 47–64.
19. Robert A. Dunstan, Smart Battery Power Availability Feature Based on Battery-Specific Characteristics, U.S. Patent 5,541,489, July 30, 1996.
20. Lawrence E. Piercey, Smart Battery System and Interface, U.S. Patent, 5,557,188, September 17, 1996.
21. Bradley A. Perkins, Smart Battery, U.S. Patent 5,747,189, May 5, 1998.
22. Wayne D. Kurle, Stephen B. Johnson, Rockland W. Nordness, Stephen I. Firman, Douglas M. Gustarson, Peter Y. Choi, Smart Battery with Maintenance and Testing Functions, Communications, and Display, U.S. Patent 6,198,253, March 6, 2001.

23. Pierre M. Combaluzier, Remote Smart Battery, 5,973,475, October 26, 1999.
24. Randall L. Hess, Patrick R. Cooper, Armando Interiano, and Joseph F. Freiman, Battery Charge Monitor and Fuel Gauge, U.S. Patent 5,315,228, May 24, 1994.
25. Robert A. Dunstan, Smart Battery Charger System, U.S. Patent 5,572,110, November 5, 1996.
26. A. H. Zimmerman, Self-Discharge Losses in Lithium-Ion Cells, *IEEE AES Systems Magazine*, February 2004, pp. 19–24.
27. William Morchin, Trip Modeling for Electric-Powered Bicycles, IEEE Technical Applications Conference, Northcon96, Seattle Washington, November 4–6, 1996, pp. 373–377.

CHAPTER 5

MOTORS AND MOTOR CONTROLLERS

Battery-powered electric motors propelled more cars in the United States than did gasoline engines during the year 1900. However, the slow-speed dc propulsion motors were heavy. Long travel range was possible only with heavy battery packs that lost their energy-storing capability in a few years. Consequently, within a decade gasoline-engine-powered cars took over the passenger car market.

Early electric bicycles were heavy and had limited travel range because they were propelled by dc motors of either the series or shunt-wound type that ran on energy stored in heavy lead–acid batteries. The variable motor speed required for bicycle propulsion was inefficiently obtained with power-wasting rheostat-type controls. Consequently, the travel distance available from a fully charged battery was very limited.

As shown in Chapter 3, available now are 3.6-V lithium ion battery cells that can store 128 Wh of energy per kilogram of weight. Assume that during each day 30 percent of the energy content of a 100-Ah cell is consumed and that the cell is recharged every night. Then 82 years after this cell went into service, it would need to be replaced with a new cell. Thus, the annual battery cost would be trivial when compared with the annual cost of an automobile-starting battery that carries only a 60-month guarantee.

Technology developed since 1990 has revolutionized the design of battery-powered motors that can propel a bicycle at speeds that the driver can command. The key development is the solid-state electronic power inverter that efficiently converts dc electric power from a battery into variable-frequency three-phase alternating current. A high-speed synchronous motor that runs on this three-phase power can propel a bicycle, through a gear train, at the speed commanded by the bicycle rider. This motor needs no field current because its field flux is supplied in its rotor by permanent magnets instead of by the heavy rotating field coils that the old dc motors contained. The current in the rotating armature of the old dc motors had to come through a speed-limiting commutator and carbon-brush assembly. This new high-speed synchronous motor weighs much less than the traditional motor. Motor design parameters are developed so that the synchronous motor and its power-supplying inverter can be designed in many configurations for meeting the performance requirements of a given bicycle configuration.

Electric Bicycles: A Guide to Design and Use, by William C. Morchin and Henry Oman

Determining requirements for a motor, selecting a battery, and the bicycle propulsion motor configuration involves evaluating many factors in an accurate manner. This subject is covered in Chapter 6.

5.1 FUNDAMENTAL PRINCIPLES OF ELECTRIC MOTORS

A current-carrying conductor in a magnetic field will have a magnetic force pushing it. This force turns the shaft of motors that receive dc power, as well as induction and synchronous motors that run on ac power. All three types of motors have been used for propelling electric bicycles. In the paragraphs that follow we show how the shafts of these motors are made to rotate and how to evaluate motor performance factors when selecting electric propulsion systems for bicycles.

A current-carrying conductor in a magnetic field is illustrated in Figure 5.1. The uniform magnetic field in the zone between north and south magnetic poles is shown in Figure 5.1a. The current-carrying conductor has around it the field shown in Figure 5.1b. Figure 5.1c shows the direction into which the conductor pushes with the force f, measured in newtons (N). This force has the following value:

$$f = \beta L I \times 10^{-5} \quad (\text{N}) \tag{5.1}$$

where β = flux density, lines/cm^2

L = length of the conductor, cm

I = strength of the current, A

As a conductor moves out of the magnetic field, the force that it contributes to making the armature revolve diminishes. Therefore, a high-performance motor must be designed to have the most possible current-carrying conductors in the magnetic field. For example, Figure 5.2 shows the torque producing elements in a dc motor that has a commutator in which each coil's ends terminate on adjacent commutator segments. One of the multiturn coils that are adjacent to the S symbol starts at one commutator segment and is wound into two grooves that are parallel to the shaft.

Figure 5.2 shows how the armature coils are connected in series, and their current flows into the series-connected coils from the commutator segment under

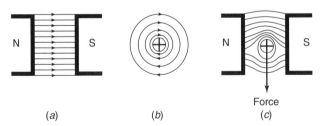

(a)   (b)   (c)

Figure 5.1   Current-carrying conductor between the poles of a magnet distorts the field between the poles and produces a force on the conductor. (From Christie [1], p. 61. Reproduced with permission of McGraw-Hill.)

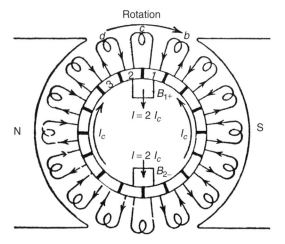

Figure 5.2   In a dc motor the stationary carbon brushes deliver current to the rotating armature windings through the commutator bars. The current flow in each coil reverses when its commutator bar passes under a brush. (From Christie [1], p. 258. Reproduced with permission of McGraw-Hill.)

the B^+ brush. The current flows out of the commutator through the segment under the B^- brush.

The coil symbols show how the coils are electrically connected. However, each coil is so wound that one-half of every turn is on one side of the armature, which can be under the south field pole. The other half is under the north pole. Thus, nearly every half-turn of every current-carrying coil is producing a force that makes this armature rotate in a clockwise direction. One groove at the instant shown is under the south pole and the other is under the north pole. Thus, nearly every half-turn of each current-carrying coil is producing a force that makes the armature of the motor rotate in a clockwise direction. The direction of current flowing through a coil reverses each time that the commutator segments to which the coil is connected pass under the B_1 and B_2 carbon brushes.

The coils on this armature are connected in series, and the current enters the rotating armature from carbon brush B_1 and returns to its source through carbon brush B_2. After the armature has rotated 90°, the coil that started at the S position arrives at brush, and there the direction of current flow in the coil is reversed. The current then produces a force that continues to rotate the armature in its clockwise direction as the coil passes under the north pole of the motor's field magnets.

Now available for bicycle motors are permanent magnets made from alloys that produce intense magnetic fields. These alloys were not available 50 years ago when big motors were manufactured for tasks such as propelling submarines with battery power and rolling steel with dc power from big motor-generator sets. Consequently, each field pole in a dc motor had to be an electromagnet in which its iron core was surrounded by a current-carrying coil. The current in the armature windings was limited by the back voltage that was generated in these windings as they passed under each field pole. Accidentally disconnecting the field coils from their current source eliminates this back voltage, so the armature proceeded to accelerate until it exploded and sent flying parts that lodged into walls of the surrounding building! This type of failure is not likely to occur in an electric bicycle where the magnetic flux in the motor fields is produced by permanent magnets rather than wound coils.

5.1.1 Brush-and-Commutator Motor

The traditional dc motor has field coils in a cylindrical steel housing. Current flowing through these coils produces a magnetic field that flows through the motor's armature winding, as shown in Figure 5.2. Current-carrying wires in the rotating armature are pushed by this field to produce torque in the output shaft. The commutator changes the direction of the current in the wires in each slot at the instant that the slot moves from the north pole's field into the south pole's field, or visa versa. Thus, the current flowing in each armature wire pushes on the magnetic field produced by the stator, making the rotor turn the mechanical load to which the motor shaft is coupled.

Each wire, as it moves through a magnetic field, produces a back voltage that is proportional to its velocity. At the motor's idle speed the back voltage equals the voltage at the motor's armature terminals. Turning the motor shaft at a speed higher than its idle speed converts the motor into a generator.

This brush-and-commutator dc machine was used by Thomas Edison to generate 120-V direct current for distribution to homes and businesses for powering incandescent light bulbs. Motors were soon developed for powering fans and appliances, and then electric streetcars, trains, and electric automobiles. The dc motor's variable-speed capability worked well for turning the rolling mills that produced paper and steel. The earliest bicycle use of battery energy for propelling a bicycle was with a dc motor that was built inside the hub of the bicycle's rear wheel. From a patent search we learned that Ogden Bolton, Jr., invented it in 1895 [3]. The motor is shown in Figure 5.3. In this embodiment the rotor is the exterior of the motor on which the wheel spokes are attached.

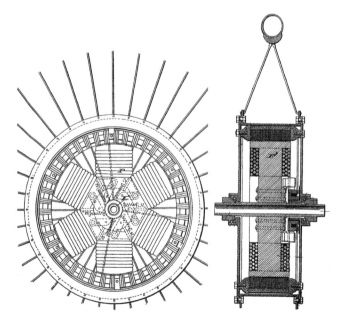

Figure 5.3 Bicycle wheel hub motor of 1895 Bolton [3].

A 1-V drop that occurs wherever a carbon brush contacts a copper surface on the commutator limits the efficiency of a brush-and-commutator dc motor. Every ampere that flows through this contact produces a 1-W loss. Also, brushes wear out, and dirt accumulating between commutator bars causes failure.

5.1.2 Induction Motors

Early in the twentieth century dc electric power distribution was abandoned because motor-generator sets would have been required for reducing transmission line voltages to a lower voltage for distributing power to users. A few commutator-type ac motors were built, but the induction motor turned out to be the most practical ac motor. This motor consists of a stator that contains windings, and a rotor in which laminations are clamped on the shaft that is held in position by bearings, as shown in Figure 5.4

The rotor of a three-phase induction motor is a simple stack of iron laminations with slots into which aluminum conductors are cast. The stator is an iron barrel containing a stack of laminations that have slots into which windings are bonded. However, the induction motor runs at slightly less than its synchronous speed, and varying the motor's speed requires varying the frequency of the supplied power.

The production of torque in a three-phase ac induction motor can be easily explained with the cross section of the motor shown in Figure 5.5, from Bose's book [2]. In it the motor's shaft supports a stack of circular disks that have slots in the rotor's outer periphery. Each slot contains a conducting bar that is usually made from aluminum. The bars are all connected together at their ends.

Alternating current flowing through the turns of stator-coil $a–a'$ in Figure 5.5 produces an oscillating magnetic field that flows through the rotor.

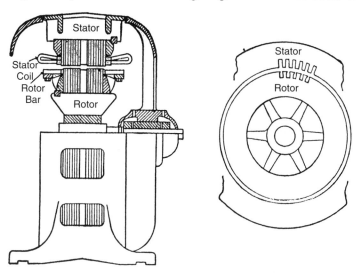

Figure 5.4   Induction motor requires no electrical connection between the stator winding and the copper bars in the rotor. (From Christie [1], p. 553. Reproduced with permission of McGraw-Hill.)

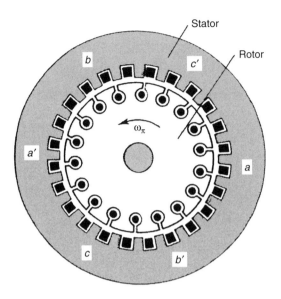

Figure 5.5 In an induction motor the current flow in the stator winding produces a rotating magnetic field that induces current flow in the rotor windings. (From Bose [2], p. 42.)

However, as the field from coil $a-a'$ reaches its positive peak value, the fields produced by coils c' and b' divert the field from coil a and return the field flux back to the stator so that it doesn't have to go all the way past the axle. A step-by-step analysis of this process shows that the three-phase winding produces a near-constant magnetic field that rotates at synchronous speed in the air gap between the rotor and stator. This rotating magnetic field induces a current in each rotor bar. This induced current produces magnetic forces that pull each rotor bar in a direction of rotation that accelerates the rotor assembly until it turns at near-synchronous speed.

An induction motor will continue to run even if only the windings in one phase are energized. However, it would not start if supplied only single-phase power. Therefore, the common single-phase motors have capacitors or resistors that deliver out-of-phase power to a starting winding in the motor.

Figure 5.6 shows an experimental adaptation of the induction motor for bicycle propulsion. In this design the conducting bars are embedded in laminated iron that forms part of the wheel rim. The windings need only to cover about 10 percent of the wheel circumference. In a sense this can be considered a linear motor.

Figure 5.6 Induction motor for propelling an electric bicycle wheel. (Photograph courtesy of Floyd A. Wyczalek, taken at Osaka, Japan, EVS-13 Electric Vehicle Conference, presentation by Klaus Hofer of the University of Bielefeld, Germany.)

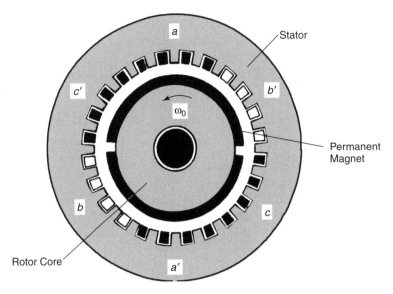

Figure 5.7   In a permanent magnet motor the stator winding produces a rotating magnetic field that pulls the permanent magnets in the rotor to make them rotate at synchronous speed. (From Bose [2], p. 63.)

5.1.3   Efficient Permanent Magnet Synchronous Motors

The rotating field produced by the three-phase winding in the air gap of an induction motor can also pull the electromagnets in a synchronous motor at its synchronous speed (Fig. 5.7). This avoids the loss of power that is dissipated in exciting magnetizing currents in the induction motor's rotor bars by forcing the rotor to turn at a subsynchronous speed. The input power that is delivered to the induction motor must include the I^2R power losses in the stator winding and rotor bars.

These losses are avoided in a synchronous motor in which the current that produces the magnetic fields on the rotor poles is delivered to the rotor's field coils through slip rings on the motor's shaft. Getting the higher efficiency of a synchronous motor requires a dc exciter on the motor shaft or a power supply that supplies the required excitation current. This made synchronous motors more expensive than induction motors and limited their applications in the past.

The important new developments in synchronous motor technology are the availability of rare-earth permanent magnets and solid-state variable-frequency power supplies. Permanent magnets produce the motor's required magnetic field flux without consuming energy. The motor then runs at a speed that corresponds precisely to the frequency of its input power that can be supplied by a battery-powered solid-state dc-to-ac inverter.

Michael Kutter designed a brushless permanent-magnet synchronous motor that he installed inside a bicycle's wheel hub [4] (Fig. 5.8). In this embodiment a set of sprocket wheels driven by a crank pedal chain are integrated with the motor. A lever actuated by the rider can engage the drive of the chain so that both

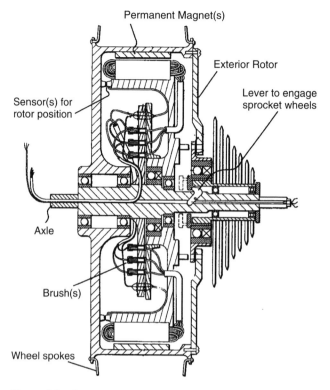

Figure 5.8   Permanent magnet dc wheel hub motor [4].

the motor and the human pedal power can be combined to propel the bicycle. The design has permanent magnets on the periphery of the rotor and uses field wires mounted to the stator. Power and sense wires reach the field wires through a hollow axle. One-way clutch bearings prevent the motor from driving the chain sprocket wheels.

The motor is direct drive and therefore has no speed reduction gears. The motor therefore must have high torque and preferably high efficiency at low speed. For instance, at 10-km/h (8 mph) a 26-inch (66-cm) bicycle wheel rotates at 80 rpm (8.4 rad/s). Motor efficiency will be low at this speed, perhaps about 30 percent as shown by Welch [5]. King-Jet Tseng and G. H. Chen [6], in their design of a hub motor for automobile applications, showed that it was possible to achieve 89 percent efficiency at a motor speed of 800 rpm. This efficiency at 800 rpm is nearly the same as that shown by Welch [5] at 800 rpm, the same reference for which the 30 percent value was shown at 80 rpm. To achieve higher motor efficiency, one must raise the operating motor speed and use gears to obtain the desired wheel speed. Planetary gears are an obvious solution.

The components of a bicycle propulsion system in which the battery supplies the power to a permanent magnet synchronous motor are shown in Figure 5.9. In Section 5.4 we show how the motor controls for battery-powered electric bicycles can be designed.

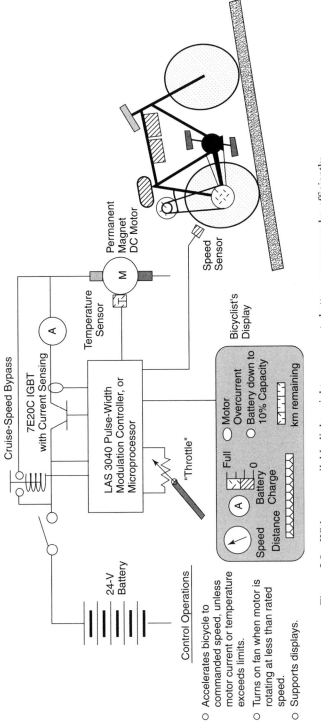

Figure 5.9   With now-available lightweight components battery power can be efficiently converted to the variable speed and torque needed for propelling bicycles. (From Binod Kumar and Henry Oman, Power Control for Electric Bicycles, IEEE NAECON Conference, 1993.)

117

5.2 MOTOR CHARACTERISTICS FOR ELECTRIC BICYCLE PROPULSION

Motor manufacturers typically describe their motor performance with data on the motor torque as a function of motor speed. They further rate their motor in terms of maximum/continuous operating torque or equivalently maximum/continuous rated power. The maximum ratings refer to short-term use of the motor for which it can safely draw current before internal temperatures buildup to damaging levels. The time intervals appropriate for this operation are measured in relatively small values of minutes. The continuous ratings are as implied: continuous operation without damage.

5.2.1 Torque–Speed Characteristics

The typical torque–speed characteristics of the motors referred to as permanent magnet, either brushless or brushed, and either ac or dc and dc shunt motors, are shown in Figure 5.10. The current that flows in the motor armature, and the magnetic flux that the armature is exposed to, is the cause for torque and motion. The motion of the armature wires within the magnetic flux creates a back electromotive force (voltage). At slow speeds the back electromotive force (emf) produced by the rotor is low and much current can flow to create high torque. At high speed the back emf produced is high and little current can flow. Low torque is created at high speeds.

   The figure illustrates three possible curves for an example motor. Each curve is for a different applied voltage. One curve can be scaled to another by the ratio of applied voltages to the motor. The higher voltage results in higher torque and higher speed. The opposite is true for lower applied voltage. This voltage scaling is sufficiently accurate for our purposes. It does, however, vary

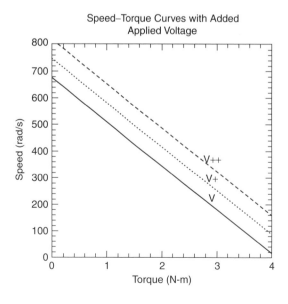

Figure 5.10   Typical torque–speed characteristics of motors referred to as permanent magnet, either brushless or brushed and shunt motors.

according to tolerances and nonlinearities within the motor components and other factors. We have found some departures from this scaling both in the offset of one curve to another and the slope of one curve relative to another. Some motors will follow the described scaling better than others will.

The torque–speed curves can be expressed as an equation for linear torque characteristics:

$$T = T_s - k_m\, \omega \qquad (5.2)$$

where T_s is the stall torque, for example, the resulting torque with a blocked motor shaft, k_m is the slope of the curve, and ω is the rotating speed. Stall torque and the torque slope are normal motor specifications provided by manufacturers.

5.2.2   Motor Output Power

Recall from Chapter 2 how power is related to force and speed. With that knowledge the output power of a motor from the formula $P = FV$, expressed in watts when F (force) is expressed in newtons (N) and V velocity is expressed in meters per second (m/s). Refer to the Appendix for conversion of force, speed, and power to other units. We have been working with translational (straight in-line) forces. Now we must deal with rotating forces. Translational force is to be converted to rotational with:

$$T = Fr \qquad (5.3)$$

where T is torque, r is the radius of the acting translational force F. Converting translational speed to rotational with $\omega = V/r$ we can solve for power in terms of T and ω to find

$$P = T\omega \quad \text{(W)} \qquad (5.4)$$

where T is in newton-meters (N-m) and ω is in terms of radians per second. The Appendix can be used to convert torque units to other values.

Motor output power can be determined in terms of the motor torque constants of stall torque and torque slope values described above:

$$P = T_s\omega - k_m\omega^2 \quad \text{(W)} \qquad (5.5)$$

Note this is a parabolic-shaped function. The power starts out at 0, increases parabolically to a maximum, and decreases parabolically to 0. Because this is a motor and it is connected to a source of electric power, the output power cannot be negative. Figure 5.11 shows the power output (the shaft power) of an example permanent magnet brushless motor. It is neither practical nor typical to operate the motor near the maximum power point. It is more typical to operate the motor near its maximum efficiency, which does not correspond to the maximum power point. And to save weight, it is more practical to make the thermal design match this same operating point.

For nonlinear speed–torque motor characteristics the above functions do not apply. Data points chosen from published or measured speed–torque data must be used to determine motor output power as a function of speed should

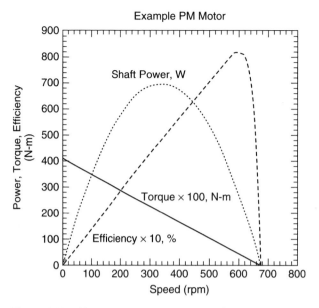

Figure 5.11   Shows the power output (shaft power) and the torque and efficiency of an example permanent magnet brushless motor as a function of angular velocity ω in rpm.

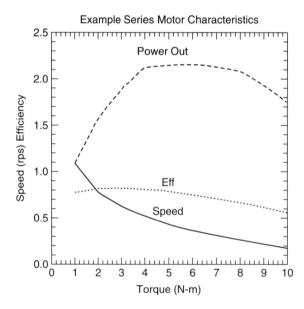

Figure 5.12   Generic series motor power, efficiency, and speed vs. torque characteristics.

such power data not be available. One example is a generic series motor as shown in Figure 5.12. We have graphically chosen sets of torque and speed data from this figure and computed the motor output power. The result is shown in Figure 5.13 along with similar data from a shunt motor. The effect of the nonlinear speed–torque characteristics on output power is obvious. Figure 5.14 compares the corresponding efficiency characteristics.

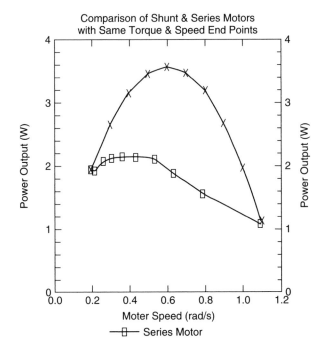

Figure 5.13
Comparison of shunt and
series motor output
powers for each motor
having the same torque
and speed end operating
points.

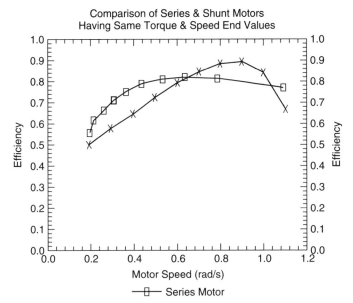

Figure 5.14   Comparison of shunt and series motor efficiency characteristics for each
motor having the same torque and speed end operating points.

5.3   GEAR RATIO DETERMINATION

Table 2.6 or Figure 2.2 gives values of power applied by the bicycle drive wheel necessary for travel under various conditions of speed, head winds, and road grades. Also in that chapter power transmission efficiencies were given. With this knowledge the motor speed and shaft power and the wheel rotation speed can be determined. This can be done mathematically or graphically. Both methods are described because only graphic information may be available on a motor or conversely only motor constants known. Alternatives are described if neither is available.

5.3.1   Mathematical Method for Determining Gear Ratio

Equation (5.5) can be solved for speed:

$$\omega_m = \frac{T_s + (T_s^2 - 4k_m P_m)^{0.5}}{2k_m} \quad \text{(rad/s)} \tag{5.6}$$

where the subscript m is used for motor power (P) and speed (ω) to distinguish these values from what will be use later for the bicycle wheel speed and power. The other terms are the same as described before. Any power value used from Table 2.6 or Figure 2.2 *should be increased* by the amount of loss expected between the motor output shaft and the bicycle wheel being powered by the motor. Some sample values are shown in Table 5.1. Dividing by one of the values or a combination of the selected efficiency values expressed as a decimal fraction will increase the motor power value by an approximate amount. Power loss is listed in the table as an alternative to the use of efficiency. These values are based on our experience of observing the additional current drain on the battery when the motor is running without a load compared to when it is driving the wheel and it is not engaged with the road surface. This loss will include additional losses not accounted for with gearing reduction alone. It will include at least wheel windage and bearing losses.

TABLE 5.1   Efficiencies (%) and Losses (W) of Transmitting Power from Motor to Wheel

Mechanism	Reduction	Efficiency	Power Loss([a])
Sprockets and chain		95	30–50
Gear box	4 to 1	92	
	10 to 1	87	
	14 to 1	83	
Planetary	3 to 1	95	30–50
Belt	7 to 1	87	40–80
Drive wheel	14 to 1	85	40–90

[a]This power loss is based on the additional current the motor draws to drive the wheel when not engaged with the road surface.

Recall that it was previously stated that the power versus speed curve was parabolically shaped. One can visualize then that there should be two values of motor speed for any given constant motor power. The reason only one is shown here is that the higher speed value corresponds to operating the motor from light loads to heavier loads. This operation is in contrast to where one would operate the motor from nearly a stalled rotor condition to a lighter load. If the motor is operated from a near-stall condition to a normal travel speed, the motor will draw a very high current that the battery must be sized to deliver. This is a reason for starting one's bicycle in motion before initiating motor power.

Now that we have an equation for motor speed, we need one for bicycle wheel speed. Taking the ratio of the two will determine the gear ratio. Finding the wheel speed is a matter of converting the translational speed of the bicycle to the rotational speed of the bicycle wheel. The conversion is

$$\omega_b = 2S/d \quad \text{(rad/s)} \tag{5.7}$$

where S is the bicycle speed and d is the bicycle wheel diameter when S and d are in the same length units and seconds is the unit of time for speed. Since d is usually expressed in centimeters or inches and S is expressed as kilometers/hour or miles/hour, we need to do some conversions to equivalent equations:

$$\omega_b = 35.2 S_{\text{mph}}/d_{\text{in}} \quad \text{(rad/s)} \tag{5.8}$$

$$\omega_b = 55.6 S_{\text{kph}} f/d_{\text{cm}} \quad \text{(rad/s)} \tag{5.9}$$

where d_{in} is wheel diameter in inches and d_{cm} is in centimeters, S_{mph} is bicycle speed in miles per hour and S_{kph} is bicycle speed in kilometers per hour. Note that measured wheel diameter can differ from the tire diameter rating. We haven't made a study of this but we have noticed a loaded 26-inch rated bicycle tire has an effective wheel diameter of 25.5 inches. The wheel radius from the axle to the road surface should be used to find the effective wheel diameter. From the Appendix you will find that the above-determined wheel speeds can be converted to revolutions/minute (rpm) by multiplying by 9.55, should motor speed be desired in rpm units.

The final result for gear ratio is $G_r = \omega_m/\omega_b$. This is the amount by which the motor speed must be reduced.

5.3.2   Graphical Method for Determining Gear Ratio

For the graphical method please refer to Figure 5.11. We will assume that you have either calculated a curve similar to this one or else you obtained one from the manufacturer. Recall our comments about the maximum operating and continuous operating power of motors. We have not shown this limit for the example motor. It should be obtained from the manufacturer.

Choose some conditions for which the gear ratio is to be determined. To illustrate, assume a starting point of no head wind and level terrain. Using the appropriate data points from Table 2.6 and reproduced in the first two columns

TABLE 5.2 Power Speed Points Chosen for Gear Ratio Selection

Speed (km/h)	Power (W)	Wheel Speed (rps)
12.9	40	11
19.3	86	16.6
25.7	161	22.1
32.2	277	27.6

of Table 5.2, we calculate wheel speed as shown in the last column of the table. The last column was calculated using 25.5 inches for the wheel diameter d. We plotted these points of power and wheel speed on the graph of Figure 5.15. Those points are the line of the leftmost squares on the graph. The motor, when connected directly to the hub without gears, would operate at a speed where a best-fit curve drawn through the squares intercepts the motor power curve. This appears to occur at a bicycle speed of 13 km/h. You will recall our remarks about operating a motor in this part of the curve: The motor will draw a very large current near its stall condition. In this example we have not increased the above power values to adjust the output power of the motor upward to account for any loss between the motor shaft and the road surface as one should do.

Now, by trial and error, we guess at a gear ratio by which we increase the above-tabulated wheel speed values. To aid in that guess, we can observe where we would like to operate the motor. It would make good sense to operate it well below its maximum power point and on the high-speed side of the parabolic power curve. We choose a gear ratio of 27:1.

The points shown by this scaling are the squares on the right-hand side of the parabolic curve. We draw a best-fit by eye-judgment line between these

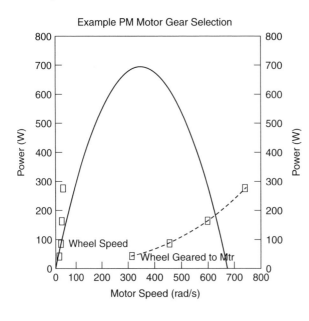

Figure 5.15 Graphical intersection of motor and road load (0 percent grade) lines with and without gearing. "Wheel speed" notation shows approximate operating point if no gearing is used. "Wheel geared to mtr" notation shows operating point at intersection when a 27:1 gear ratio is used.

points. Projecting the intersection with the motor power curve downward, we get an operating motor speed of about 625 rps. We see that this intersecting point is about 20 percent of the span above the 25.7-km/h point between 25.7 and 32.2 km/h. The bicycle will travel at about $25.7 + 0.2 \times (32.2 - 25.7)$ km/h or 27 km/h with a motor output of 185 W.

We observe that this operating point is a bit high and away from where the peak efficiency is expected. A lower gear ratio would be chosen to lower the motor speed. The result, however, would be a faster travel speed and possibly more power use. One would have to determine if efficiency is increasing faster than the motor output power to know if power use actually increases. Furthermore, recall that this result is for travel on a level surface.

Propulsion up an incline with this same example gear ratio may result in too large of a load on the motor. Figure 5.16 shows a new operating point with a 27:1 gear ratio. The bicycle will now travel at about 22 km/h with a motor output of 510 W. The motor's ability to dissipate heat, and on how close this point is to the peak efficiency, will determine if this can be a satisfactory operating point. Higher road inclines can make the results more marginal. With a higher gear ratio the power consumption can be decreased. A lower travel speed will be the result. For a gear ratio increased to 41:1, the required power output of the motor is reduced to 360 W and the travel speed reduced to 17 km/h on the 6 percent incline.

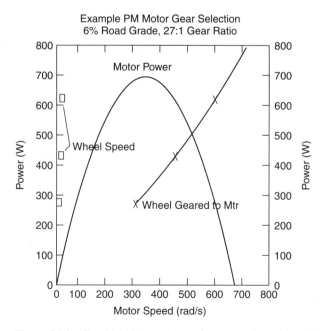

Figure 5.16 Graphical intersection of motor and road load (6 percent grade) lines with and without gearing. "Wheel speed" notation shows approximate operating point if no gearing is used. "Wheel geared to mtr" notation shows operating point at intersection when a 27:1 gear ratio is used.

Thorough communication with the motor manufacturer is required for optimization of motor use on the electric bicycle. It is possible to use a lower power motor with selectable gear ratios rather than a higher power motor with no gear ratio selection.

5.3.3 Using Nameplate Data for Determination of Gear Ratio

Unfortunately, this may be the paragraph you use the most if you are motorizing your own bicycle. Short of measuring the output power of a motor, we will provide you some guidance. Hopefully, this will save you from going to the effort of putting an unsuitable motor along with gearing onto you bicycle.

You may be fortunate enough to find a motor with a nameplate that specifies voltage, current, and speed. If so, multiplying current and voltage will give you the continuous operating input power. Assuming efficiency in the neighborhood of 65 percent will approximate the output power. An alternative to assuming efficiency is to estimate the maximum efficiency with measurements. You can refine your efficiency assumption if you can measure the no-load current of the motor and the line input resistance of the motor. Assuming that you do, use the following formula to determine the maximum efficiency of the motor:

$$\text{Eff} = [1 - (I_{\text{no-load}}/I_{\text{start}})^{0.5}]^2 \times 100\% \qquad (5.10)$$

where $I_{\text{no-load}}$ is the no-load current and I_{start} is the starting current. We recommend that you calculate the starting current by first measuring the line input resistance of the motor and dividing that into the voltage you will use to test the motor. The starting current can be measured by using a voltage lower than the motor operating voltage and gradually increasing it until the armature just begins to rotate. However, if instead, the horsepower (hp) rating is given, then that is the operating output power of the motor. The nameplate speed value is the no-load speed of the motor.

To determine the operating speed of the motor, we found that maximum efficiency is usually at approximately $\frac{1}{7}$th of stall current. This corresponds to a speed of $\frac{6}{7}$th of the no-load or nameplate speed, of the motor. Higher efficiencies than the usual 65 to 75 percent will occur at this speed. The efficiency function will show up as a concave curve peaked at $\frac{6}{7}$th of the speed range with a sharp drop to zero at the no-load speed and a gradual somewhat linear decrease to zero near zero speed. Figure 5.11 illustrates this. With this shape in mind and observing the parabolic power output curve, you can see that maximum output power, which occurs at 50 percent of no-load speed, and maximum efficiency of the motor, does not occur at the same speed.

The operating point of the motor is normally between the speeds for maximum power and maximum efficiency and closer to the maximum efficiency speed. The average speed between the two points is about two-thirds of the no-load speed. We believe that a good estimate for motor speed would be above this value and close to $\frac{5}{7}$th of the no-load speed.

Now, to find the gear ratio, use $\frac{5}{7}$th of the no-load speed for the motor and assume a gear ratio, G_r. Calculate a wheel speed $\omega_b = G_r \times \omega_m$. Using the above formulas for bicycle wheel speed, solve for the bicycle speed:

$$S = d_{cm}\ \omega_b / 55.6 \quad \text{(km/h)} \quad \text{or} \quad S = d_{in}\ \omega_b / 35.2 \quad \text{(mph)} \qquad (5.11)$$

when ω_b is in radians/second. Convert to radians/second from revolutions/minute by multiplying by 9.55. With the combined motor power and bicycle speed values in hand, refer to Table 2.6 or Figure 2.2 to find a close combination of similar values. If that combination of values corresponds to the desired travel conditions; for example, road grade and wind speed, you have selected the right gear ratio. If not, you will have to select another gear ratio and try again until you are satisfied with the road grade and wind speed travel conditions.

One reminder: If you operate your motor at a voltage different from the nameplate voltage, then the speed values and torque values should be scaled as described in Section 5.2.1.

5.4   MOTOR CONTROL

Controlling the motor is an obvious necessity when riding an electric bicycle. Initiating and changing the electric current that flows in the motor controls a motor. In some motors, as we have described, electric current is also used to create the magnetic field within the motor. Strengthening or weakening this field controls the speed and power of the motor. However, an open field circuit in a shunt motor can result in a motor speed runaway problem. The problem will be alleviated by the battery limitation of the maximum current and fusing.

The technology of controlling motors has advanced so much that motor design is being relieved of past constraints. However, the controlling circuits can be so complex that products of specialty companies must be used. Design of this part of electric bicycles is left to system engineering. One must coordinate and allocate equipment requirements between motor manufacturers and electronic control manufacturers. In this section we can present the basics to further your understanding.

We first discuss dc brush motor control from basic switching to use of semiconductor switching. That is followed by a discussion of brushless motor control of single- and three-phase motors as an introduction to the control of multiphase motors.

5.4.1   Brush Motor Control

Power is switched into the windings of a dc motor through a commutator and brushes. Brush wear can be of concern to the bicyclist. Consider lifetime use of the bicycle to be 20,000 miles. For an average speed of 12 mph using a motor, one would accumulate 1700 h of operation on the motor. This time span may or may not be a significant part of the motor's life. If adequately designed and sized for the loads, this time span should be insignificant with inspection and maintenance. Brush wear is dependent on the application and control circuitry. Simple on/off switching can cause sparking at the brush–commutator interface due to emf energy stored in the motor coils. Spark suppression circuitry, to be discussed below, will enhance the brush life in these situations.

Controlling the current flowing to the motor can control motor speed. The simplest method is to use an on/off switch. For this situation travel speed is dictated solely by the motor and load. A step up from this would be to select one of two battery voltages with a switch. We found this a satisfactory solution that has the virtue of simplicity yet can provide reasonable control. We could pedal at a very leisurely pace, turn on the low battery voltage, and travel in the 9-mph range or turn on the higher battery voltage and travel in the 14-mph range. The next step while maintaining simplicity is to use a rheostat (a variable resistor) to control motor current. Though simple, this method wastes battery energy. The next step up in complexity is to use a pulsed motor current drive. The pulses are electronically controlled and of very short duration relative to the inertial characteristics of the motor. This type of controller is referred to as a pulse width modulator (PWM) controller.

Pulse width modulators use power semiconductors that are switched on and off by square-wave oscillators. The duty cycle of on to off is controlled by a low-power rheostat set by the bicyclist. Power semiconductors of the insulated gate bipolar transistor (IGBT) variety or metal–oxide semiconductor field-effect transistors(MOSFETs) for low voltages are used. These devices have been developed to the point of having no more series resistance than a normal toggle switch.

The power semiconductor can be placed circuitwise between the battery positive terminal and the motor or between the negative terminal and the motor. When placed to switch the positive supply voltage, it is referred to as high-side switching. It is referred to as low-side switching when the negative supply voltage is switched. Figure 5.17 illustrates the idea. However, should there be a failure of the high-side power switch from output to ground, the motor will stop. This type of power semiconductor failure is dominant. However, for a low-side switch, this kind of failure will cause the motor to run until mechanically switched off.

Higher voltage driver devices must be used for high-side switching unless the power semiconductor's drain and source are isolated from the substrate material. This adds complexity to the control and sensing circuits within the controller.

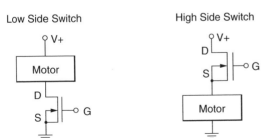

Figure 5.17   Low-side and high-side switch motor control.

The semiconductor power switch will have to withstand the inductive back emf of the motor. Adequate snubber circuits are required to protect the semiconductor device. While durable for rated loads, the semiconductor devices are particularly sensitive to high-voltage transients. While a 100-V rated device can suffice for the current load of a motor, it cannot survive the common 600-V or more transient voltages generated by switching on and off a motor. This transient is caused by the energy stored in the magnetic field about the current-carrying wires in the motor. The energy in joules (J) is given by the equation:

$$E = 0.5IL^2 \quad \text{(J)} \tag{5.12}$$

where I is the electric current in amperes and L is the inductance of current circuit in henries. When the current is reduced, the collapsing magnetic field generates a voltage that tries to keep the current flowing in the same direction. The quicker the current is turned off, the higher the voltage. Transistors can turn off in a matter of tens of nanoseconds. In one of our tests a 4-A current into a 0.4-mH load generated 25 V that tried to keep current flowing through a MOSFET. This was in addition to the 12 V from the battery.

The snubber typically consists of a capacitor in series with a resistor to absorb the sharp voltage rise from a motor back emf transient. A fast diode is placed across the resistor to permit the current due to the transient voltage to flow into the capacitor. The resistor (with shunting diode)–capacitor pair is connected in parallel with the motor. The capacitor then releases its stored energy into the resistor and battery or other loads. In addition a fast response bucking diode is connected in parallel with the power semiconductor. A bucking diode is likely to be built into the IGBT device. It is a diode connected for nonconduction under normal circumstances. It conducts under much higher voltages as can occur in the transient turn-off conditions, and short-circuits the transient voltage around the power semiconductor. As an example we found that a resistor of 2 Ω and a capacitor of 0.47 μF along with a fast-acting zener diode with a zener voltage of 36 V reduced the above-mentioned transient to insignificance. The sample electric bicycle motor with an inductance of 0.4 mH and a winding resistance of 0.17 Ω ran with a supply voltage of 24 V dc and a current of 8.5 A. The reader is referred to Undeland [7] for detailed design guidance for other motor conditions.

An example high-side driver circuit (without a snubber) of a controller for a brush dc motor is shown in Figure 5.18. The example solves a problem of gate control voltage that has to be larger than the supply voltage to turn-on the final MOSFET [8]. V_{in} is an oscillating voltage with polarities as shown in Table 5.3. The duty cycle of this voltage is controllable by the bicyclist by means of a potentiometer. Low-duty cycles cause relatively small average motor current, and high-duty factors have the opposite effect. The semiconductors shown (Q_1 through Q_3) are MOSFETs. They act as switches under the control of the voltage placed on the gate of the transistor. The on/off state of these switches follows the patterns shown in Table 5.3. The transistor Q_3 is sized to handle the motor currents and voltages. It is normally "on" and is turned "off" by a negative

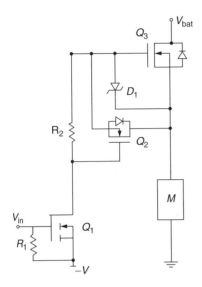

Figure 5.18   High-side PWM controller for a brushed dc motor [8].

TABLE 5.3   Switch States and Voltages for High-Side Motor Driver

V_{in}	Q_1	Q_2	Q_3	V_{motor}
$-V$	Off	On	On	$(V+) - (V_{q3})$
0	On	Off	Off	0

voltage at its gate. The negative voltage is controlled by transistor Q_1. A manual switch can replace Q_1 if one desires to use manual control of the motor. This manual switch can be a small low-power microswitch mounted near one of the handlebar grips. Q_2 provides for rapid turn-on transition of Q_3. The resistors R_1 and R_2 provide current flow between the gate and source for Q_1 and Q_2. The zener diode D limits the "off" bias voltage applied to the gate of Q_3.

5.4.2   Brushless Motor Control

Eliminating brushes improves reliability but at the expense of more complexity in the controls of the brushless dc (BLDC) motor. In addition, the resulting brushless motor is more complex, as already discussed. By eliminating brushes, the BLDC motor improves reliability, eliminates arcing and dust residue, and reduces maintenance. Other advantages are less rotor inertia and more efficient heat dissipation. Greater heat dissipation permits more current and therefore greater torque for a given maximum allowable temperature.

A brief comparison between a system using a dc motor versus one using a BLDC motor is shown in Table 5.4. All electronic means of motor control will be susceptible to moisture. So a bicyclists must keep in mind what electronics are being exposed to water or mud.

TABLE 5.4   Relative Cost Comparison Between DC and BLDC Motor Systems

Feature	BLDC	DC
Motor cost	2	1
Control cost	3.1	1
System cost	2.5	1

Cost values are relative to 1.

Single-Phase Brushless Motor Control   Referring to Figure 5.19 we see an arrangement of high- and low-side switches appearing as an H shape. The switches are controlled such that Q_1 and Q_4 are turned on while Q_2 and Q_3 are turned off. That creates a motor current traveling to the right in the diagram. A short time later, these two pairs of switches are set to the opposite settings. This causes the current in the motor to flow in the opposite direction, to the left in the diagram. In this manner the current is made to alternate so that an inverter has been created to change the battery dc voltage to an ac voltage.

With proper control of the switching frequency a rotating magnetic field can cause a brushless single-phase motor to run. Rotor position sensors provide the information to control circuits for setting this switching frequency. One example of this sensing, shown in Figure 5.19, uses low-resistance current sensing resistors in series with motor coils and subsequent amplifiers. Other means are often used, such as Hall magnetic field sensors mounted within the motor or motor back-emf sensing. All of these methods require either an initial motor movement or

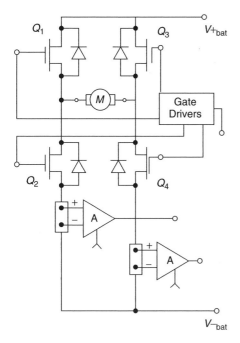

Figure 5.19   H-drive network for controlling a dc motor. The H drive consists of a high-side driver with a complimentary low-side driver such that when Q_1 receives a turn-on signal Q_2 will receive a turn-off signal.

a fortunate starting position of the rotor within the initial magnetic field. The later is the usual case. However, the startup can be relatively slow, requiring the adjustment of the switching frequency to match the inertia of the motor. Once running, the speed of the motor can be controlled by changing the switching frequency and/or the magnitude of the current.

The average magnitude of the current is changed by a PWM square-wave oscillator that controls the duty factor of the driving frequency waveform. The PWM circuits in turn receive setting commands from the bicyclist speed control or the pedal-force sensing system. The PWM commands are sent to the gate drivers shown in Figure 5.19. The circuits within the gate driver block are similar to those shown in the high-side switch circuit (Fig. 5.18). There is a multitude of these gate driver circuits within the block. A high-side driver will have a complimentary low-side driver such that when Q_1 receives a turn-on signal Q_2 receives a turn-off signal. Although these functions can be performed with individual components, they are often performed within specialized integrated circuits. The motor drive switching transistors can also be integrated together with the gate drivers if the motor power and voltage values are low enough or if the state of the art for handling such values is advanced enough for electric bicycle use.

Multiphase Brushless Motor Control Motors of more than one phase can be controlled in a fashion similar to single-phase motors. Figure 5.20 shows a motor drive used for either two- or three-phase motor control. An extra "leg" has been added to the H-drive circuit of Figure 5.19. An additional leg would be added for each added motor phase. Additional phases make it possible for greater motor efficiency.

The motor control is performed in a manner similar to that for the single-phase motor. Referring to Figure 5.20, note that when a turn-on signal is at either a, b, or c of the motor drive transistor gate there must be a turn-off signal at a', b', or c'. For example, let Q_1 and Q_4 be turned-on as denoted by a and b'

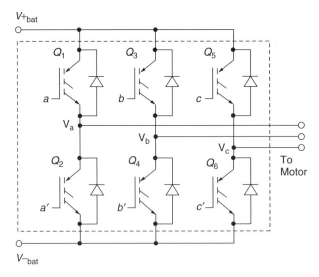

Figure 5.20 Example of a driving circuit for a three-phase motor.

TABLE 5.5 Motor Controller Switch Signals and Phase Voltages

Gate Signals			Motor Phase Voltage		
a	b	c	V_{ab}	V_{bc}	V_{ca}
1	0	0	1	0	−1
1	1	0	0	1	−1
0	1	0	−1	1	0
0	1	1	−1	0	1
0	0	1	0	−1	1
1	0	1	1	−1	0

having a signal of 1 and the complimentary set a' and b denoted by a signal of 0 with Q_2 and Q_3 being turned off. In this situation the battery voltage will be applied between the two wires labeled V_a and V_b. This is referred to as a winding-to-winding voltage of V_{ab}. Depending upon the position of the motor coils, the line-to-line voltages must be either positive or negative relative to one another so that the resulting torque of the phase windings reinforces each other. Table 5.5 illustrates the timing and phasing of the control signals and the resultant phase voltage indications [9]. For this illustration a value of 1 for V_{ab} means a positive battery voltage is applied at V_a. For −1 a negative voltage is applied. For 0, no voltage is applied. For the three-phase motor six of the eight possible settings are used. The table shows only a, b, and c signals because a', b', and c' are equal and opposite.

Figure 5.21 shows representation of the line-to-line phase voltages. There are six states of voltage combinations: one for each phase and each direction of current flow in each phase. The voltages are applied so that (1) first current flows

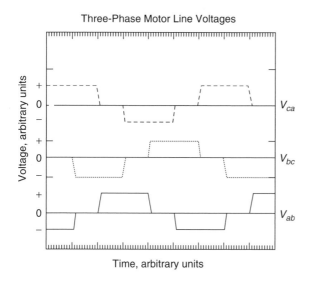

Figure 5.21 Timing diagram for three-phase motor line-to-line voltages.

into phase A and out of B and C, then (2) out of C from A and B, then (3) into B and out of A and C, then (4) out of A from B and C, then (5) into C and out of A and B, and finally (6) out of B from A and C. The process repeats. The dead time of each phase can be used to sense the phase back emf for controlling the motor.

Simulation Tools The design of motor controls using semiconductor micro-circuits is greatly assisted by simulation computer programs. The nonlinearities and complexities of control of motors can be overwhelming without such a tool unless one is very experienced. Use of the software tools can significantly shorten a design cycle [10].

With simulation computer programs one can model a motor and its load as well as the motor control circuit. This is especially useful for multiphase motors driven by inverters that convert the battery supply voltage to multiple phases of alternating current.

5.4.3 Pedal Torque Used to Control Motor Power

Regulations of some countries have spurred the development of sophisticated motor control mechanisms. The requirements impose operations so that the bicycle can not run on motor-only power and also further complicate operation so that no motor power is applied above a certain speed. An example of supplementing human pedal force with torque produced by a bicycle-propulsion motor is shown in Figure 5.22. This has made it necessary for manufacturers to design sensing systems to determine speed of the bicycle and when human force is applied to the pedals. The signals from these sensing systems are used to control the motor power to be applied for propulsion.

The control circuit determines the ratio of human power and motor power by (1) finding the bicycle speed and (2) finding the human power with a pedal torque measurement. These two values are used as input to a motor controller. Torque is measured by measuring the angle of displacement between two rotors or disks connected between the pedal crank and the drive chain sprocket wheel. The two rotors or disks are separated by a spring mechanism for which the displacement varies in proportion to human force on the crank pedals. The deformation causes

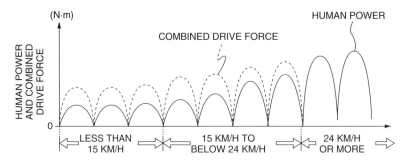

Figure 5.22   Motor and human power combinations for propelling an electric bicycle [11].

a relative angle change between the two rotors or disks. A semiconductor Hall device is used for angular detection. The Hall device has an applied voltage that causes change in an output voltage in proportion to magnetic flux lines passing through the device. A small magnet attached to one rotor or disk and a Hall device attached to the other rotor or disk will cause a magnetic field to vary in proportion to the angular change between the devices.

The control signals are used in such a way that if no pedal force is detected the motor will not run. With pedal force applied, the motor power is controlled so that with greater pedal force more motor power is applied. One way clutches are used is between hollow concentric shafts that transmit pedal power and motor power to the drive chain sprocket wheel. The one-way clutches prevent the motor from driving the pedal cranks and preventing the pedal cranks from driving the motor. For bicycle speeds less than 15 km/h the two forces are equal. Between

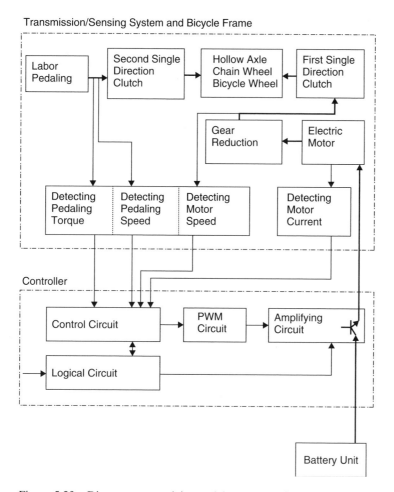

Figure 5.23   Diagram summarizing pedal torque sensing and motor control [12].

bicycle speed of 15 and 24 km/h the motor force is linearly decreased until no motor force is applied for speeds greater than 24 km/h.

Another embodiment of using pedal torque sensing is given by Chin-Yu Chao et al. [12]. They include sensing of motor speed. All three signals are sent to a microprocessor for which the motor power assistance to the peddler is determined. Their objective is the same as that of Takada et al. [11]. This implementation is shown in Figure 5.23.

The idea of using one-way clutches is illustrated by us in Figure 5.24. In this illustration torque sensing is not incorporated. The figure shows another means of isolating the pedal crank force from the motor force and permitting both forces to be combined for propelling the bicycle.

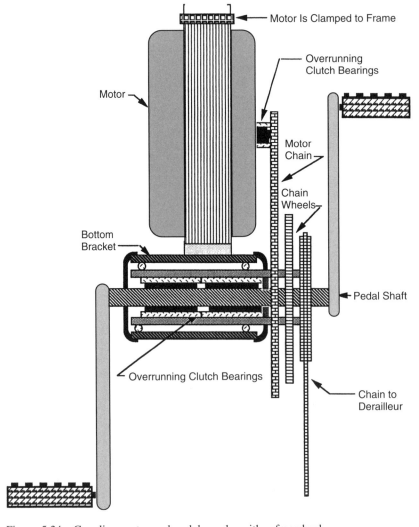

Figure 5.24 Coupling motor and pedal cranks with a freewheel.

REFERENCES

1. Clarence V. Christie, *The Theory and Characteristics of Electrical Circuits and Machinery*, McGraw-Hill, New York, 1931.
2. Bimal K. Bose, *Power Electronics and Variable Frequency Drives*, IEEE Press, Piscatawdy, NJ 2002.
3. Ogden Bolton, Jr., Electrical Bicycle, U.S. Patent 552,271, December 31, 1895.
4. Michael Kutter, Hybrid Drive Mechanism for a Vehicle Driven by Muscle Power, with an Auxiliary Electric Motor, U.S. Patent 6,286,616, September 11, 2001.
5. Richard H. Welch, Jr., Improving the Power Efficiency of Electric Motors—Part 2: Steady State Power Efficiency, *PCIM*, April 1998, pp. 22–28.
6. King-Jet Tseng, and G. H. Chen, Computer-Aided Design and Analysis of Direct-Driven Wheel Motor Drive, *IEEE Transactions on Power Electronics*, Vol. 12, No. 3, May 1997.
7. Mohan N. Undeland, and William Robbins, *Power Electronics: Converters, Applications, and Design*, Wiley, New York, 1995.
8. Richard A. Blanchard, High-Side Switch with Depletion-Mode Device, U.S. Patent 6,538,279, March 25, 2003.
9. Issa Panahi, et al., Generate Advanced PWM Signals Using DSPs, *Electronic Design*, May 1, 1998, pp. 83–90.
10. Powersim Inc., PSIM Simulation for Power Electronics and Motor Control, Advertising Flyer, 2004.
11. Yutaka Takada, Hiroshi Miyazawa, Akihito Hetake, Kaniaki Tanaka, Hirosli Nakazato, and Katsami Shikai, Sensor, Drive Force Auxiliary Device, U.S. Patent 6,163,148, December 19, 2000.
12. Chin-Yu Chao, Chih-Jin Wang, Yin-Jao Luo, and Yuh-Wen Hwang, Power Transmission and Pedal Force Sensing System for an Electric Bicycle, U.S. Patent 6,196,347, March 6, 2001.

CHAPTER **6**

THE SYSTEM DESIGN

Systems engineering is a new aerospace technology for making a thorough design of complex transportation systems. We show that systems engineering principles can be applied to electric bicycle design.

6.1 INTRODUCTION

In the past costly transportation system failures have occurred. For example, ferries and cargo ships propelled by efficient gas turbines, rather than steam turbines, were once built. However, their unpredicted defect turned out to be maintenance services, which required skilled mechanics who were working at airports and not generally available at docks.

At the beginning of the twentieth century more electric cars than gasoline cars were running in the United States. The electric cars gradually disappeared because their batteries lost their energy storage capacity after a few years and required costly replacement. Electric vehicle production soon ceased. However, a skilled systems engineering analysis would have revealed that the Edison nickel–iron battery would have solved this problem. In 1998 an Edison nickel–iron battery, recovered from a forklift truck in a junk yard, was found to still function satisfactorily.

The word *system* means many things in many fields. To a physician, it means a part of the body. To an electrochemist, it means an electricity-producing or consuming reaction. In public-utility work, it means an electrically integrated power-handling entity. The word *system design* in "electric bicycle system design" is the process of selecting the best electric bicycle for a defined application. "Best" means the design that produces a bicycle with lowest lifetime cost and satisfies the users' needs.

Systems engineering was developed in the aerospace field during World War II. An example of a recent success is the battery-powered Mars Rover vehicle that had to first survive the severe acceleration and vibration during launch into Earth orbit and then acceleration into interplanetary orbit. After a period of zero-gravity environment came the deceleration into Mars orbit, followed by a shock of landing on the surface of Mars. The batteries and propulsion motors then performed as commanded by a sophisticated control that communicated with personnel on Earth.

Electric Bicycles: A Guide to Design and Use, by William C. Morchin and Henry Oman
Copyright © 2006 The Institute of Electrical and Electronics Engineers, Inc.

In the sections that follow we show how systems engineering can be used for establishing pertinent requirements for an electric bicycle and then designing it and selecting its components. We then describe in detail the systems engineering procedure, which involves (1) establishing clear requirements, (2) creating and evaluating preliminary designs, as well as the final design, (3) evaluating test results of power-delivering components to determine if design changes will produce better vehicle lifetime performance, and (4) thorough testing of the prototype electric bicycle. The key evaluation criteria are the electric bicycle's performance and its life-cycle cost.

6.2 SETTING UP THE ELECTRIC BICYCLE SYSTEMS DESIGN

System engineering, which is required in the conceptualization and construction of complex systems, is an important part of any major project. System engineering starts in the initial stages of a project and continues throughout the project design, thereby having a great impact on the successful completion of the project. Many computer tools have been designed to aid the system designer. One word that is the key to understanding these tools is *system*. Webster's generic definition describes a system as "a regularly interacting or independent group of items forming a unified whole." In an engineering sense, a system is an assembly of components, some of which may have to be modified or redesigned to meet the system's design objectives. "Systems" range from extremely high level concepts, such as "a motorized vehicle for individuals," to the low level, such as "a drive train."

6.2.1 System Design Project Functions

The project functions performed by the designer of a major component of a system are shown in Figure 6.1. He begins with a concept definition stage, which broadly defines the goals of the project. For example, the designer of permanent magnet propulsion motors for a new electric bicycle model may be asked to deliver 25 percent more torque than a previous motor delivered in a low-speed range. The designer would determine from his records and analysis that a new rotor design using bigger magnets would deliver the requested torque. He then makes a functional design that his shop uses to build and test a prototype motor.

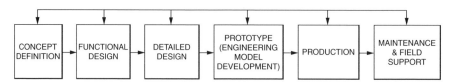

Figure 6.1 Block diagram of project functions performed by a designer of a system component. (After M. T. Talbott, H. L. Burks, R. W. Shaw, D. D. Strasburg, and K. K. Hutchison, Method and Apparatus for System Design, U.S. Patent 5,355,317, October 11, 1994.)

After a motor being tested delivers the required torque, the designer can release revised drawings and specifications that enable a factory to begin producing the higher torque motors.

After production is started, maintenance and field support is provided. Note that there is continuous feedback from the prototype production and maintenance blocks, which is incorporated in the concept definition, functional design, and detailed design. This feedback is used to correct assumptions, mistakes, and improve understandings to enhance and achieve a better design.

6.2.2 System Design Cycle

An engineer developing a new system works in an environment that is much more complex than the component designer's environment. This more complex system design cycle is outlined in Figure 6.2

Inputs The system engineering cycle begins with the identification and evaluation of the operational needs of the system being considered. An example of the dynamic environment, which the system designer must quickly evaluate, occurred in the year 2004 after Floyd Wyczalek had predicted that the world's oil production would reach its peak in 2005; see Chapter 3. Retail prices of gasoline and fuel rose by over 50 percent. Our cities are already heavily polluted with gas fumes, and the aging population cannot or will not ride on pedaled bicycles. This environment has produced a need for electric bicycles in other nations. China's fast growing annual production of "standard" electric bicycles had passed 275,000 per year [1]. In his analysis the system design engineer who is developing electric bicycles needs data on the interests of commuters at train stations, pedestrians, bicyclists riding on bicycle paths, and other potential users. Examples of useful data for planning an electric bicycle development program include:

What is the acceptable price range of an electric bicycle?

What is the longest acceptable commuting time to your workplace?

Can you recharge batteries at work?

Under what weather conditions would you not ride an electric bicycle?

Additional pertinent data can be obtained from traffic engineers, police headquarters, bicycle shops, and chambers of commerce for developing the performance requirements of electric bicycles in a community. Pertinent information can be obtained by attending several bicycle and electric bicycle exhibitions as well as conferences on batteries and motors. You can browse the Web. Laws and regulations that might affect the system are important. Standards that are applicable to the system need to be studied. The observation of actions of pedestrians and commuters can be applied. All these facts are interpreted and prioritized in the next step of concept definition. This interpretation may be validated with the potential customer through interviews. After that the system requirements are defined.

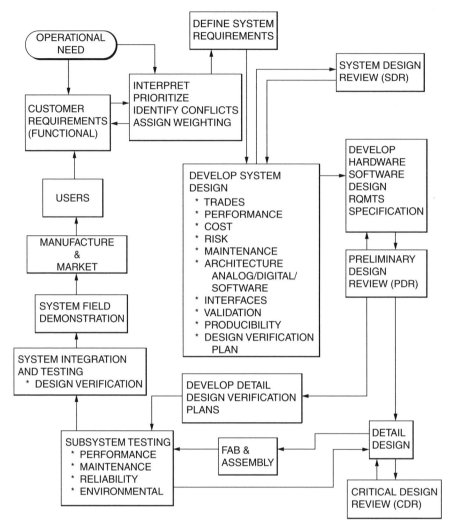

Figure 6.2   Diagram of a typical system design cycle. (After M. T. Talbott, H. L. Burks, R. W. Shaw, D. D. Strasburg, and K. K. Hutchison, Method and Apparatus for System Design, U.S. Patent 5,355,317, October 11, 1994.)

The defined system requirements are used to develop three categories of system design requirements: real, derived, and assumed. Some requirements are easy to quantify, for example, regulations imposed by some governments (see Chapter 1). Requirements that cannot be determined precisely become assumptions. All important assumptions need to be identified with their limits and flagged for perturbation in the electric bicycle system design process.

Process   A key to systems design is the consideration of all reasonable alternatives. Requirements are quantified with nominal and possible extreme values. Alternative ways of meeting the requirements need consideration. The first step is to configure preliminary designs to the depth necessary for cost estimating and defining interfaces. The developed system design includes design trade-offs, estimated performance, estimated costs, risks, maintenance plans, validation, and producibility.

Life-cycle costs need to include alternative possibilities that meet the requirements. These need to be ranked with other pertinent factors. Perturb the requirements to see what happens. Consider upper and lower limits of pertinent factors.

The computation of life-cycle cost is a straightforward procedure. Life-cycle cost is the total cost of a process or project pertinent to the owner. However, owners have differing interests. An example of differing interests are owners of bicycles for sports applications versus commuting. Effective systems design requires computation of life-cycle cost of alternative approaches as well as perturbations of inputs to the process. Using a computer to produce an electronic spreadsheet program speeds the process. Ranking the life-cycle costs is a first step in electric bicycle systems design evaluation. Some alternatives will be obvious rejects, and others will be so nearly equal that further optimization and perturbations of inputs to establish sensitivities are not needed.

The system design includes an architecture in which the system is broken into logical groups. For example, Figure 6.3 shows the architecture of an electric bicycle broken down into mechanical and electrical subsystems, and the components in each subsystem are shown. Interfaces between components must also be defined. The information generated in the system design is then reviewed. The purpose of the review is to answer this question: Is the design a profitable cost-effective solution that will satisfy the customer? This interactive process in system design is refined until an acceptable design has been generated.

After an acceptable design is generated, a design requirements specification is generated. The specification would include, among all other items, a description of the propulsion motor in terms of its cost in pertinent quantities, its performance, and physical attributes. With these data a motor manufacturer could provide a quote. The design requirements specification is reviewed in a preliminary design review to verify that the design has been adequately and thoroughly described so

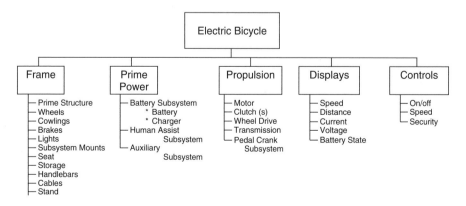

Figure 6.3   Subsystem components of an electric bicycle.

that the design can proceed to its detail design stage. Development expense and safety need to be examined carefully when new apparatus is being considered. Refining of the design and specifications continues until the final design review.

After the preliminary design review, a detailed design is developed. A design verification plan is concurrently developed. This plan includes a review for adequacy. Refinement of the detailed design continues until it is accepted at the critical design review. After this acceptance, a detailed design of the electric bicycle is delivered to a shop for fabrication and assembly. The resulting vehicle is tested for performance, maintainability, reliability, and durability for survival in its operating environmental. Information from the system testing is used to refine the detailed design if necessary.

Output   Once testing is complete and system integration is verified, the electric bicycle is demonstrated to verify that the system requirements and operational needs have been met and that the customer will be satisfied. The system is then marketed and manufactured. During its use, customer reaction is sought for further refining the design.

6.2.3   Importance of Preserving Data During Design Process

As the cycle progresses from the concept stage to the design stage, numerous requirements often become apparent. All persons concerned with the design should record these requirements for review. This will assist in maintaining a consistent set of requirements for the project. Throughout the process, the system designer will need to continue to refine information, break down the requirements, and define the functional structure of the solution. System-level requirements need to flow downward in the hierarchy to ensure that the delivered system meets its requirements. Likewise, implementation details need to flow upward in the hierarchy in order to allow the system designer to further refine the definition of the functional structure of the solution.

The most important contribution of the electric bicycle's systems design is the perturbing of requirements and evaluation of the results. With the support of modern computers, the electric bicycle systems designer can quickly evaluate many alternatives and focus on the best choices. An important element to consider is the different types of support tools that are appropriate at different system levels. Current software [e.g., computer-aided software engineering (CASE)] and hardware [e.g., computer-aided engineering (CAE)] design tools are directed toward efficiency of use in a specific environment, rather than over the broad range of system levels. As such, several different tools are necessary at the different system levels, which complicates communication of necessary information between system levels. Furthermore, the specific tools place different operational requirements on the user, thereby reducing the efficiency of using the programs.

6.2.4   User-Pertinent Data in System Design Reviews

Much of the data evaluated in design reviews is pertinent to the bicycle user's cost and attainable benefits, which may not always be clear to the user. An example is an ordinary lead–acid motorcycle ignition battery, which might appeal to the user because of its low cost. If charged with an ordinary 12-V charger, this battery will last only around 50 charge–discharge cycles if each trip consumes 70 percent of the battery's energy capacity. Consequently, after every 50 trips the battery will need to be replaced. This is the same phenomenon that ended the use of electric cars in the early 1900s.

In a design review batteries are evaluated on the basis of the electric bicycle's life-cycle cost and the per-kilometer cost of travel. The alternative to the low-cost motorcycle battery is the nickel–cadmium battery that is used in communication satellites and lasts for thousands of charge–discharge cycles. Its depth of discharge and charge control are optimized in systems engineering analyses. In the sections that follow we illustrate simple system engineering techniques that can give the lowest life-cycle and per-kilometer traveled costs for specific travel requirements.

6.2.5 Uses of Systems Engineering Procedure to Evaluate New Opportunities

The coming shortage of petroleum for vehicle propulsion will motivate investigation into more efficient means of travel, including electric bicycles. Availability of a proven systems engineering program and the data that it contains will make quick evaluation of new technologies and opportunities.

For example, solar-powered transportation has been frequently tested but has not been adopted because no solar energy is available when the sun is obscured by clouds and during nights. However, there is a new product, the zinc–air fuel cell that could make solar energy practical for electric bicycles. This fuel cell (see Chapter 3) has a membrane with zinc powder on one side and an air electrode on the other side. The cell is filled with potassium hydroxide electrolyte.

6.3 SOME EXAMPLES OF BICYCLE–SYSTEM TRADES

6.3.1 Trading System Weight with Efficiency

What if one could obtain an additional 10 percent in system efficiency by adding 3 lb (1.4 kg) for a higher efficiency motor? Answering this question requires a study to trade weight with efficiency. This is part of the process of developing a system design through the use of trade studies. Our approach to finding a solution is to determine the relationship at a system level between vehicle weight and system efficiency for constant values of energy use. Thus, for a constant outcome, we will know how much system efficiency will have to be improved for a given amount of weight increase. This answer can be a guide to creating a motor design.

We analytically determine the power required to propel an electric bicycle with a computer model for various weight, system efficiencies, and road grades. The object of the modeling is to construct a table of energy consumption values for a range of weight and efficiency values. Constructing contours of constant values through the matrix of energy use values provides the gradient of efficiency change for weight change.

We concluded that for each 5 kg of weight change there must be a 1 percent change in system efficiency to maintain constant energy consumption performance for a 0 percent travel surface grade. This result and others for the other road grades are shown in Table 6.1.

The values for the frequency of occurrence of road grades is given to judge the impact of various road grades. The frequencies of occurrence values are summarized for grade intervals of ± 0.5 percent grade for each grade value in the table. We conclude that for each kilogram of motor weight increase the efficiency must be increased 0.3 percent. This conclusion is reached by assuming the electric-powered bicycle should be operated at peak efficiency for road grades up to 1.5 percent. This example is meant as a guide to one such trade study that can be used to optimize performance.

6.3.2   Comparison of Specific Efficiencies of Electric Vehicles

The preceding section showed that for every 3- to 5-kg change in the electric bicycle weight there is a corresponding change of 1 percent in the bicycle efficiency in its use of energy from the battery. In this section we compare the electric bicycle with other vehicles. A common measure for the comparison is how much rider mass is moved per unit of energy per hour, which is referred to as specific efficiency expressed in kilogram/Watt-hour/kilometer.

One can determine from Table 1.1 that the energy use is equal to about 9.5 Wh/km, the average travel speed is 23 km/h, and the specific energy consumption is 15.6 kg/Wh/km (including the bicycle mass and assuming a 80-kg rider).

Typically, an electric bicycle will use about 10 Wh for each kilometer traveled and with rider mass of about 80 kg, resulting in a specific efficiency of 8 kg/Wh/km. A point of comparison is an electric-powered motorcycle [5]. This 676-kg motorcycle, when traveling 32 km/h had a specific efficiency of 15.7 kg/Wh/km for two 80-kg riders. At 64 km/h it was 7 kg/Wh/km. For an electric-powered automobile carrying four 80-kg passengers, the specific efficiency is about 2 kg/Wh/km. The hybrid electric automobile assuming equivalent 60-mpg gasoline consumption has a specific efficiency of 1.1. The electric bicycle is the most efficient of the four modes of travel:

TABLE 6.1   Efficiency Sensitivity to System Weight

Road Grade (%)	Efficiency Sensitivity (kg/%)	Grade Frequency of Occurrence (%)
0	5	50
1.5	3.3	32
3	2.3	10

- Electric bicycle: 8 kg/Wh/km (25 km/h, 1 rider)
- Electric motorcycle: 7 (54 km/h, 2 riders)
- Electric automobile: 2 (4 riders at 160 Wh/km)
- Hybrid automobile: 1.1 (4 riders, 60-mpg equivalent gasoline consumption)

6.3.3   Travel Speed Selection

The use of the bicycle to achieve the longest travel distance with the available battery energy is another example of the application of the system approach. There is a range of choices to be made in how fast to travel and how much human power to apply to the pedal cranks. The two extremes are to either use no human power or use no motor power. Morchin [2] has shown us that it is possible to use calculus to find the recommended travel speed for various surface grades that results in minimum energy use for a selected travel time over a given distance. His example showed that the following speeds should be selected after having chosen an average speed to travel:

Level surface: travel 2 percent higher than the average

2 percent grades: travel 7 percent below average

3 percent grades: travel 11 percent below average

4 percent grades: travel 25 percent below average

Morchin assumed all motor power and no human assistance. He used a motor efficiency that varied with road grade; 67.1 percent for 0 percent grades, 67.5 percent for 2 percent grades, 64 percent for 3 percent grades, and 59 percent for 4 percent road grades. The distribution of road grades was similar to those in Table 6.1. Travel speeds would be different if the distribution were different. In addition, if motor efficiency varied differently for various motor loads, the recommended speeds would also be different. The limited study showed that it is possible to analytically optimize travel. This is a subject that requires further development.

6.3.4   Applying Battery Ragone Plots

The Ragone plot shown in Figure 3.4, if graphed in log–log form, will permit additional interpretation. This form of the Ragone plot is shown in Figure 6.4. Here we have drawn the lines showing time rates of battery use to full discharge. Constant rates of use of 0.5, 1 and 2 h are shown. Intersections of these lines with the battery characteristics define the regions of interest for electric bicycle travel distance applications. Although longer rated battery use can be applicable, they would require carrying more battery weight.

Here too, we can see the usefulness of the ultracapacitor for supplying the extra power needed during acceleration or short hill climbs. The average-man data and gasoline engine data are added for perspective.

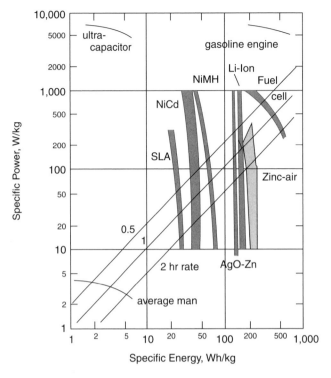

Figure 6.4   Specific power and energy characteristics compared for various energy sources. (Battery data from TMF^{tm} Technology—The Answer for High Power Battery Requirements, white paper by Bolder Technologies Corp., January 15, 1996. Ultra-capacitor and gasoline engine data from *Electric and Hybrid Vehicles: Design Fundamentals*, Figure 3.17 p. 66, 2003. Fuel cell data from Figure 3.4, Chapter 3. Average-man data from authors.)

6.3.5   Using Ultracapacitors to Reduce Battery Size

The ultracapacitor, also referred to as a supercapacitor, consists of organic elec-trolyte immersed in a space between plates of graphitelike microcrystalline carbon layers. These capacitors have very high capacity for storing electrostatic charge, on the order of 15 to 20 F/cm^3 at voltages of 2.5 to 4 V.

We can evaluate the benefit of using the high-power capacity of an ultra-capacitor to reduce the power drain on the battery during hill climbing and accelerating periods of bicycle travel. Hill climbing is the most power demand-ing. For instance, from Chapter 2 we see that for about 25 km/h travel on level roads a power of 161 W (at 100 percent efficiency) is required, similarly about 271 W for a 6 percent grade at half speed (100 percent efficiency). Considering efficiency, which on the level might be 85 percent and on the 6 percent grade is likely to decrease to, let's assume 65 percent. Thus, without human assistance the power required on the level is 190 W and on the 6 percent grade it is 417 W. The difference to be supplied by the ultracapacitor is 227 W. Figure 6.4 shows that the best energy density is 8 Wh/kg for an ultracapacitor with a power density of about 7000 W/kg.

For this example we will see that the ultracapacitor is energy limited, if we assume the 6 percent grade is 0.5 km long when traveling 10 km/h. The travel time will be 0.05 h, thus requiring 7 Wh (0.05 h × 227 W). A 1-kg ultracapacitor would be needed to supply this energy, whereas that same capacitor could deliver 8 kW. With the cyclist supplementing the motor and ultracapacitor with 100 W of human power, the 6 percent grade hill could be nearly 1 km long for the same capacitor ratings.

Experimental capacitors have been shown to have 27 to 60 Wh/kg specific energy ratings [3, 4]. These ratings would make them competitive with batteries. In fact, Honda Motor Company has combined such new capacitors with fuel cells in vehicles now being leased to the city of Los Angeles [3].

Charging the ultracapacitor is also an issue. Niiori et al. [4] show data indicating capacitor-charging times of about 4000 s (67 min). In Chapter 4 we showed that the recoverable energy during coasting down a 4 percent grade hill would be about 4 Wh/km. If that were done at a 10 km/h rate, we would only gain 6 min of charge. We would need 10 such downhill segments to obtain capacitor hill-climbing assistance for one uphill travel segment. Other charging means such as the battery or fuel cell is a practical necessity.

The slow speed travel with electric bicycles appears to negate the use of ultracapacitors if only battery charging of the capacitor is available. The capacitor may help with accelerating, but the cyclist is not likely to be stressed in accelerating the bicycle. This example shows that the system designer can trade between power and energy sharing between the cyclist, the battery, and the possibility of an ultracapacitor for short hill climbs. Perhaps future developments will encourage the use of ultracapacitors in reducing the high motor currents during uphill travel.

6.4   SYSTEM DESIGN EXAMPLE

Shown in Figure 6.2 are the steps that a bicycle design team would take in developing a successful electric bicycle. For a design example a brief summary is presented of a small number of all the possible system design paths. In the example only the essential steps of motor and battery selection are covered. The frame design, the controller, battery charger specifics, and specific cost values are not addressed.

6.4.1   Customer Requirements

A survey has indicated a desired maximum travel speed of 25 km/h on level road surfaces for a maximum electric-powered propulsion distance of 40 km. Furthermore, the bicycle should be capable of having an 8-year battery life and that the battery should be recharged while parked. The bicycle will also be used for a 10-km daily commute to work (20 km/day for 6 days/week). These are examples of the technical aspects of what a customer desires. Bicycle cost to the customer is critical but beyond the scope of our example.

6.4.2   Interpret, Prioritize, Identify, and Assign Weighting

The following points are derived from the preliminary set of customer requirements:

- A distance of 40 km implies that at 10 Wh/km a battery delivers energy of 400 Wh. For a 70 percent depth of discharge that gives a battery rating of 571 Wh.
- A commute of 20 km means 200 Wh are used each day.
- Three battery recharges are needed each week. For 8-year battery life that means battery life must be 3 charges/week $\times$ 52 weeks/year $\times$ 8 years = 1248 cycles.
- A speed of 25 km/h for an 80-kg rider (implying total system weight of 100 to 120 kg) means a motor power of about 165 W for level travel or 185 W in a 10-km/h head wind going up a 3 percent slope at a speed of 13 km/h (Table 2.6). These values are for 100 percent efficiency. It will be assumed that the rider is willing to add 20 W of human power to the motor when traveling up the 3 percent grade. Thus, the power applied to the road will be 165 W. No human power is assumed for traveling on the level grade. Assuming 35 W of gearing efficiency loss, the motor output power must be 200 W.

6.4.3   Define System Requirements

The first step in defining system requirements is to find the combination of propulsion motor type and battery type that can meet the preliminary set of customer requirements. A preliminary set of requirements comes from the inferences made in Section 6.4.2. These requirements can be later developed into system design steps shown in Figure 6.1, which the system designer must subsequently perform. However, the detailed performance of the motor and battery are needed in the development of the system design cycle in which alternatives are compared in a manner that produces useful results.

The lightest motor for starting the analysis could be a high-speed permanent magnet three-phase ac motor that is coupled by a reduction gear to an overrunning clutch on the shaft that can also be turned by the rider's pedals. It should be capable of delivering at least 200 W of output power. Performance data for the motor is available from manufacturers. Also available are dc-to-ac variable-frequency converters that can convert the dc battery power into three-phase power for the propulsion motor. The controller should have a steady current rating of at least 10 A and a voltage rating equal to or higher than the maximum allowed battery voltage for rider safety. Perhaps 48 V can be assumed at this point in the design cycle. A battery that can meet the propulsion power and energy and weight limitations will need to await a battery trade study. And it is too early in this stage of the design cycle to specify weight limitations.

The motor power rating can be calculated from the speed required for climbing the steepest hill without rider assistance to be a 2 percent grade. The required battery capacity can be estimated from the customer requirements, motor efficiency, and maximum depths of discharge that can be allowed and still meet the operating lifetime requirement. For this example a deliverable battery energy rating of 400 Wh will be used and a power rating commensurate with the motor power required for hill climbing or accelerating at 1 m/s^2 will be assumed.

6.4.4 Develop System Design

With the above data, plus data and quotations from manufacturers, a preliminary design can be configured for entering into the model shown in Figure 6.2. Alternate configurations can be developed and entered into this model. The system engineering analysis will produce data such as the costs of the alternatives, their risks, and future developments that can result in better performance of the battery-powered electric bicycle. The system analysis can then evaluate design changes that will adapt the basic electric bicycle to serve the needs of other applications. Trade studies to identify motor and battery characteristics follow.

Motor Selection Trade Studies Motor representative data are shown in Figure 6.5 for three types of electric bicycle motors. Motor data from manufacturers or suppliers can be obtained in such varying formats. The designer must then apply these data as in Chapter 2 to each motor with appropriate selection of gearing for operation near the motor peak operating efficiency. It is assumed that this has been done for each motor type and that motor B has been selected.

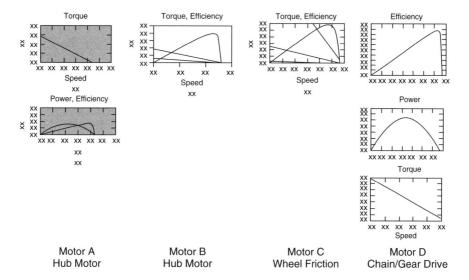

Figure 6.5   Motor representative data for three types of electric bicycle motors.

Motor B, the hub motor, is selected to investigate further. Figure 6.6 shows the example road load data for 0 and 2 percent road grades superimposed onto the manufacturers power output and motor current data. The manufacturer has several varieties within the same series of motors. Figure 6.7 shows the results of applying the road load to each motor of the series. The values to the left of each point identify the motor first by its voltage rating and then by its version number. The values to the right of each data point show the wheel size selected for the investigation.

There is only one motor and wheel size combination that meets or exceeds the desired speed. However, a very close second, the 24V4 motor mounted in a 26-inch wheel provides for about 24.5 km/h at about two-thirds the energy use. It would be appropriate to return to the motor manufacturer to request a new version with a design change to obtain the 1 percent increase in speed. Before proceeding, the motor performance must be determined for travel up hills. It is found that for a 2 percent grade, 320 W of motor output power is required.

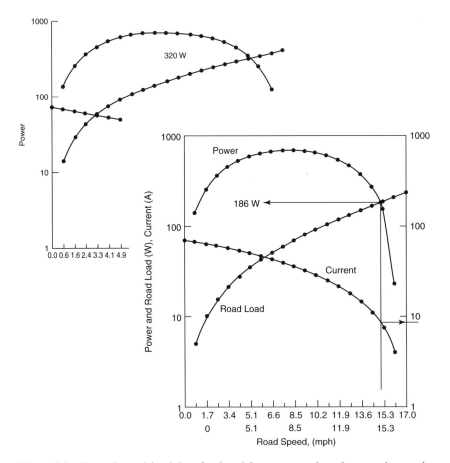

Figure 6.6 Example road load data for 0 and 2 percent road grades superimposed onto example power output and motor current data.

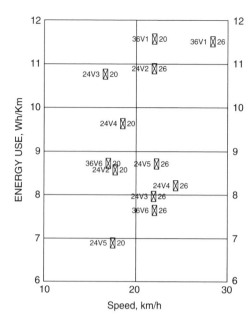

Figure 6.7 Results of applying the road load to each motor of the various 24- and 36-V models. The values to the left of each point identify the motor first by its voltage rating and then by its version number. The values to the right of each data point show the wheel size selected for the investigation.

The motor rating has an output power of 375 W to limit operating temperature below 150°C. The designer is now faced with making choices of motor control. The choices are between (1) turning off the motor, (2) depending upon the rider to sense a motor overload for slopes greater than 2 percent for which the rider would then add pedal power until the motor did not appear to be overloaded, and (3) using a pedal force torque sensing and control system that would enable the rider and motor to share the power required to travel up hills.

Battery Selection Trade Studies   A bicycle battery could have an acceptable life if a small percent of its capacity is discharged on each trip. However, this requires carrying the weight of unused battery capacity on the electric bicycle.

What is needed for a trade study of batteries is a relationship between depth of discharge and the necessary life of 1250 cycles for the battery. Such a relationship exists for the NiCd battery [Eq. (3.2)]. However, an extensive set of tests have not been published for other battery types. For these other types assumptions will be necessary. For the purpose of this trade study example, it will be assumed that the battery life changes at the same rate with depth of discharge as the NiCd shown in Figure 3.2, but with an offset established by specific life data points that are available for the other batteries.

Recent tests have shown lifetimes of over 17,000 charge–discharge cycles at a 30 percent depth of discharge in each cycle for the lithium ion battery. Lead acid batteries lose all of their energy storage capacity in less than 100 charge–discharge cycles if they are completely discharged each time. Equation (3.2) is used but with a revised exponent for each of these batteries. Rather than the exponent constant of 4.7, a value of 4.94 is used for lithium ion and 4.38 for the sealed lead–acid battery. The NiMH life characteristics are assumed to be the same as NiCd.

Using Eq. (3.2) for the NiCd battery one can predict a life of 1250 cycles if discharged no more than 65 percent. This battery must then contain the energy of 400 Wh/0.65 = 615 Wh. For a discharge rate of about 0.8C (a 1.25-h discharge rate), a battery specific mass of 48 Wh/kg is to be expected; see Figure 6.4. Therefore a 13-kg NiCd battery will be necessary. This procedure is repeated for each battery type using the above assumed exponent constants for the other battery types. Table 6.2 shows the results.

TABLE 6.2 Battery Trade Study Example

Battery Type	Depth of Discharge[a] (%)	Battery Energy (Wh)	Energy Density (Wh/kg)[b]	Battery Mass (kg)[c]	Energy Vol Density (Wh/L)[b]	Battery Volume (liters)
SLA[d]	45	889	28	31.7	70	12.7
NiCd	65	615	48	12.8	140	4.4
NiMh	65[e]	615	70	8.8	235	2.6
Li Ion	75	533	175	3.0	250	2.1
Zinc–air	70	571	250	2.3	200	2.9

	Preliminary Size[f] (cm)			Relative Cost	Power Density (W/kg)[g]	Peak Draw (W)
	Height	Width	Length			
SLA[d]	11.7	23.3	46.6	1	130	4127
NiCd	8.2	16.4	32.8	3.5	150	1923
NiMh	6.9	13.8	27.6	4.1	200	1758
Li Ion	6.4	12.9	25.7	5.6	600	1829
Zinc–air	7.1	14.2	28.4	6.9[i]	200	457

[a]To achieve at least 1250 life cycles.
[b]At a 0.8C discharge rate.
[c]To obtain the deliverable energy of 400 Wh.
[d]Sealed lead–acid.
[e]NiMH assumed to have the same cycle life characteristics as NiCd.
[f]Based on 1-, 2-, and 4-dimensional ratio.
[g]Based on maximum capability for battery weight.
[h]Scaled upward because cannot achieve 1250 cycle life. Scaled by 1250/200.

6.4.5 Develop Hardware and Software Requirements Specification

Motor Specifications   It is best to speak to motor manufacturers or suppliers for procurement of a motor that best meets the system design. As explained in Chapter 5, motors are described in terms of torque, output power, and efficiency as a function of motor shaft speed. Power and bicycle speed are the bicyclist's parameters. These parameters can be related to torque and shaft speed as explained in Chapter 5. Torque can be expressed as a function of the power delivered to the wheel, wheel diameter, and travel speed with the relationship, repeated from Chapter 2:

$$T = Pd/55.63\ S \quad \text{(N-m)} \tag{6.1}$$

where P is the power in watts, d is the wheel diameter in centimeters, and S is the travel speed in kilometers/hour. Figure 6.8 shows a general solution to this equation in the form of a nomograph. With this graph one can determine one value of the four with knowledge of the other three. The above hub motor example is used to determine the desired motor torque for the 2 percent grade conditions. The answer of about 12 N-m is shown for a motor output power of 320 W, a 66-cm wheel diameter, and a 25-km/h travel speed. Figures 6.9 and 6.10 show more specific results for two common wheel diameters.

The wheel rotation rate is given by Eq. (5.9):

$$\omega = 55.63S/d_{\text{cm}} \quad \text{(rad/s)}$$

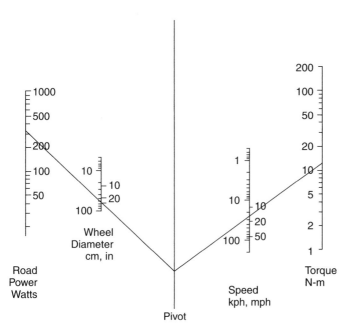

Figure 6.8   Nomograph general solution for the conversion between road power and torque. With this graph one can determine one value with knowledge of the other three. Straight lines are drawn between points on the scaled axis on each side of the pivot line intersection as illustrated. For example, a line drawn between a road power of 320 W and a wheel diameter of 66 cm results in an intersection on the pivot line. Another straight line drawn between a wheel torque of 12 N-m and that intersection on the pivot line gives an answer for the vehicle speed of 25 km/h.

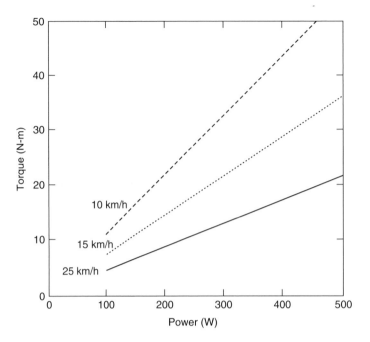

Figure 6.9 Converting power and speed to torque for a 51-cm (20-inch) wheel. Same parameter conditions as noted in Figure 6.8.

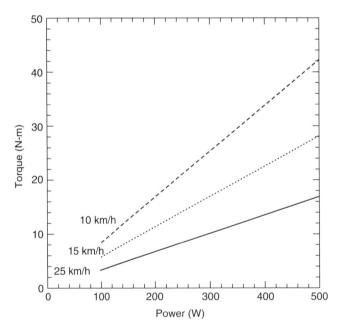

Figure 6.10 Converting power and speed to torque for a 66-cm (26-inch) wheel. Same parameter conditions as noted in Figure 6.8.

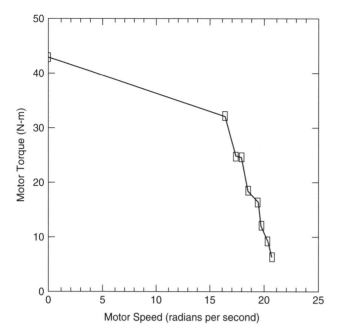

Figure 6.11   Example motor torque–speed specification.

The procedure is repeated for various road loads to obtain the results shown in Figure 6.11. The eight data points show results for road grades 0 through 6 percent in the 16-through 21-rps range. These results combine the investigation into the use of a 51-cm and a 66-cm wheel diameter. The point for 0 rps (the stall torque) is based upon the maximum current the motor is allowed to draw and/or the acceleration desired. Chapter 2 referred to a value of 43 N-m for an acceleration of 1 m/s^2. That value is carried forward to this example and is shown in Figure 6.11.

   At this point in the system design it would be beneficial to achieving a well-balanced design to contact motor manufacturers and discuss the preliminary motor specifications. Their experience with motor characteristics can be used to enhance the specifications. These enhancements combined with further rounds of bicycle performance predictions emphasizing more extreme performance parameters such as larger road grades, longer times at particular grades, greater or lesser accelerations, and use of regenerative battery charging or not. Decisions made using these results would then be fed back to other topics in the system design with a consequent change in motor and other component specifications depending upon mutually accepted compromises in each.

Battery Specifications   Battery specifications can be developed in concert with development of specifications for overall desired performance and the motor specifications. The electric bicycle success or failure rests mostly on the performance and cost of the battery. We have given numerous examples of the twentieth century failures of electric vehicles because of the short travel range.

Table 6.2 data provide the basis for creating the battery specifications. The trial battery dimensions, using a 1 : 2 : 4 size distribution, have been fit onto a trial bicycle design and found acceptable. As we found in Chapter 2, the motor current for the 36-V motor example was 32 A corresponding to a power draw from the battery of 1152 W. We see that without increasing the size of the zinc–air battery sized in Table 6.2 this battery would not be able to deliver this peak power. It can be seen that its mass would need to be increased by a factor of 2.5 to 5.8 kg.

The cycle life of the rechargeable zinc–air battery is limited to about 200 cycles in the range 70 to 100 percent of depth of discharge. There are no clear values for cycle life for lower depth of discharge. Therefore, this battery would need to be changed 6.25 times more frequently compared to the other batteries and does not meet the customer-required lifetime of 8 years. However, as pointed out in Chapter 3, some zinc–air batteries can be made such that only the zinc powder is replaced. In this sense, then, that battery would be discharged to nearly 100 percent each time, and it is used more like a "gas tank." There are no published cost data to support a cost estimate for this feature.

The sealed lead–acid battery cannot be discharged to depths similar to the NiCd, NiMH, and Li ion batteries for the same required cycle life. Consequently, to achieve the same cycle life, excess battery mass must be carried by the bicycle. Its mass for the same life is 2.5 times more than the next battery, NiCd, in Table 6.2. It appears at this preliminary stage that NiCd, NiMH, and Li ion batteries are the contenders for selection. The competition is a typical trade between size and mass being inversely proportional to cost.

At this point it would be a benefit to discuss the trade-offs between the various factors with battery manufacturers to obtain the best trade between size, cost, and performance.

6.4.6   Develop Detail Design Verification Plans

Motor Design Verification Plan   The motor design verification plan consists of, first, a dimensional and weight measurements to verify that it matches the drawings. Then, measurement of the electrical characteristics such as no-load current draw and electrical resistance would follow. Finally, measurement, as outlined in Chapter 7, of the mechanical output power will provide final verification that the motor performs to the requirements specified. Departures from that specified will have to be fed back to the design process to determine if acceptable. The verification plans are submitted for approval from the design team, including the suppliers.

Battery Verification Plan   The battery verification plan should follow the same ideas presented for the motor; first, a dimensional and weight measurement followed by measurements to determine the power and energy as specified is provided by the battery. The life-cycle characteristics of the battery might be best determined at the battery plant.

6.4.7   Subsystem Testing

Many electric bicycles that were designed and sold to the public have disappeared from the marketplace. The usual cause was a component failure that could have been corrected in the bicycle's design and test process.

After selecting a candidate motor, we perform tests to determine if indeed the motor will perform to the specifications. Motor output power test results obtained using the Prony brake method, described in Chapter 7, are shown in Figure 6.12. Should the data spread about the expected output powers be unacceptable, then a better test method needs to be investigated. A thoroughly calibrated dynamotor test set at the motor manufacturer can be a better alternative. Witnessing such testing and approving the results obtained at the manufacturing plant can be a more cost-effective solution.

6.4.8   System Integration and Testing

Rather than taking the prototype bicycle outdoors to test, an indoor testing facility might be considered here. Perhaps a test stand fixture could be used that would include a wheel dynamotor that could simulate varying motor loads, a concurrent battery testing apparatus, a concurrent display, and a control system test. Water and weather effects should be considered at this point.

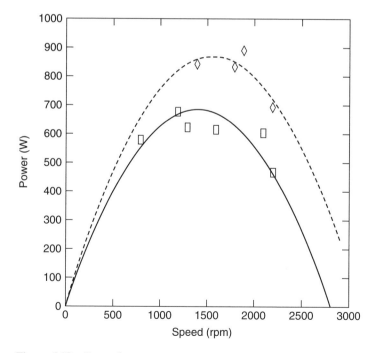

Figure 6.12   Example motor power output measurements compared to theoretical values.

6.4.9   System Field Demonstration

Ultimately, the performance of the bicycle will need to be demonstrated in a real-world environment. Designers should also consider year-long sample testing with various riders to learn about long-term effects not noticeable in the more immediate test results.

REFERENCES

1. Lester R. Brown, M. Renner, and B. Halweil, *Vital Signs 2000—The Environmental Trends That Are Shaping Our Future*, W.W. Norton, New York.
2. W. C. Morchin, Energy Management in Hybrid Electric Vehicles, 17th Digital Avionics Systems Conference, October 31–November 6, 1998, Bellevue Washington.
3. Michio Okamura, Introducing the Nanogate Capacitor, IEEE Power Electronics Society Newsletter, First Quarter, 2004.
4. Yusuke Niiori, Hiroyuki Katsukawa, Hitoshi Yoshida, Makoto Takeuchi, and Michio Okamura, Electric Double Layer Capacitor and Method for Producing the Same, U.S. Patent 6,487,066, November 26, 2002.
5. H. I. Mohamed-Nour et al., *Design Considerations in an Efficient Electric Motorcycle*, 0-7803-3631-3/97 IEEE, 1997, pp. 283–287.

CHAPTER 7

MEASUREMENT OF PERFORMANCE

To some extent whether you are an electric bicycle user, a new owner, a vendor, a planner, or designer, you will have an interest in testing and measuring some aspect of the electric bicycle. We explain here the principles involved in measuring the major performance parameters of interest. First, the measurement of propulsion efficiency of the bicycle will be explained so that you can understand how far the battery will power the bicycle. Knowledge of motor power and its efficiency is also covered. Included is the measurement of battery capacity. We discuss methods of gathering the data to assist you in selecting what method to use.

7.1 MEASURING PROPULSION POWER TO DETERMINE PROPULSION EFFICIENCY

Chapter 2 presented the analytical way of determining propulsion power. Here we present the empirical method. The measurements can be used to refine some of the constants used in the analytical method. For example, we originally used an aerodynamic drag coefficient of 1.4 for a bicycle and bicyclist. We found that a coefficient of 1.0 produces power consumptions that better correspond to values measured during our tests.

The key to our testing was the measured speed of the bicycle and rider when coasting down a constant slope on a calm day. Then we used the motor to propel the bicycle and rider up hill at the same speed. We measured the slope with a surveyor's level. In the absence of wind the bicycle reached an ultimate speed at which the windage-and-friction drag exactly balances the force from the forward component of gravity, Figure 7.1. Then we return uphill, using just enough power to make the bicycle move at the same velocity that it had coasted downhill. The required uphill power is obtained from two required equal forces, F_d and F_{wf} shown in Figure 7.1. One is the force that overcomes windage and friction at the travel speed, F_{wf}, and the other is the force that moves the bicycle and rider mass uphill, F_d. That is, $F_p = F_{wf} + F_d$.

Note that the downhill coasting test $F_{wf} = F_d = F_g \sin \varphi = F_g G$, where G is the grade expressed as a fraction, and the angle of the grade is φ. Therefore,

Electric Bicycles: A Guide to Design and Use, by William C. Morchin and Henry Oman
Copyright © 2006 The Institute of Electrical and Electronics Engineers, Inc.

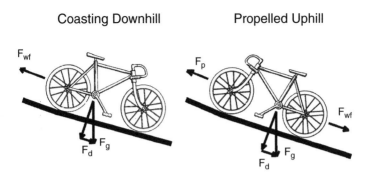

Figure 7.1   Forces involved in measurement of propulsion power.

after substituting F_d for F_{wf}, we find that $F_p = 2F_{wf}$. Thus, the power delivered by the driving wheel of the bicycle in uphill travel is twice the power required to overcome the wind and friction and is $P_p = F_p V_p$, where V_p is the constant coasting speed.

As an example, assume that a bicycle with rider weighs 122 kg (269 lb) coasting down a constant slope grade with a slope of 1.78 degrees (3.0 percent grade) reaches an ultimate speed of 5.5 m/s (20 km/h) (12 mi/h). Force F_d for this case is $122 \times 0.03 = 3.66$ kg, which is $9.807 \times 3.66 = 36$ N (newtons). See the Appendix for the conversion factor. The propulsion power is $P_p = 2F_d V_p = 2 \times 36$ N $\times 5.5$ m/s $= 396$ W.

The bicycle's battery must deliver the wheel power plus electrical and mechanical losses. The above test method cancels out the effects of some of the mechanical losses such as rolling resistance. However, the propulsion losses, such as those in the wheel drive and within the motor, are not canceled out. We can express the effect of these losses as a propulsion efficiency, which we think of as system efficiency. This system efficiency is equal to $P_b/E_p I_p$, where P_b is the power determined from the above test. The voltage, E_p, and current values, I_p, are those recorded during the travel uphill.

The system efficiency can be further identified from components of motor efficiency, electrical system efficiency, and drive system efficiency. Separate measurements or analyses can evaluate each component for the purpose of understanding what is critical to improving performance. Motor efficiency can be measured in a separate test or determined from the manufacturer's data. Circuit losses can be calculated from data presented in wire tables. Switching losses, either mechanical or solid state, can be determined from manufacturer's data. The connector losses can likewise be found in reference manuals.

7.1.1   Sources of Error

Every measurement has an error associated with it. In this case every term that describes the forces and powers involved in electric bicycle propulsion are not precise. We can only hope to accept what is reasonable. Fortunately, in the field of travel we do not need to be absolutely precise because there are many other major factors that affect the cost of travel. It seems to us that an ultimate error of 5 percent should be acceptable. If a power of 250 W is determined to be acceptable for bicycle propulsion, an error within 13 W is reasonable. We define error here as the rms (root mean square) value, which corresponds to one standard deviation from the average value of Gaussian distributed errors. This should be the case for our situation where we have a large collection of error sources with one exception. The exception is air windage effects. For all other measurements it appears to us that they can be combined in an rms fashion. The total rms error is the square root of the sum of the squares of each component error.

Note that propulsion force required for overcoming air resistance varies as the square of the bicycle's relative speed with respect to its head wind. The propulsion power delivered where the tire contacts the ground varies directly with the speed of travel over the ground. For example, a bicycle traveling at 10 m/s into a 2-m/s head wind could require 290 W of propulsion power. The same bicycle traveling at 7 m/s into a 5-m/s head wind would need only 203 W. Both cyclists would feel the same 12-m/s air in their faces as they cycle.

An important source of error could be a 1-m/s air velocity change, which is hard to measure during a test run because the local wind velocity may not be constant over the route of a test run. For example, a bicyclist riding at 10 m/s in perfectly calm air could need 198 W of propulsion power. Assume that he maintains his 10-m/s speed but enters a 1-m/s head wind. Then his power consumption rises to 243 W. Repeated tests and careful data analysis are required to reduce the effect of error sources. Statistical analysis and judgment combined with comparisons to theoretical values will assist in identifying "outlying" data to be ignored, that is, those data that are far from the norm.

Although we mention a 5 percent total error, we do not mean to imply carelessness in measurement. Theoretically, if there are 10 components of which each one has an equal error, the allowable error for each is about one-third of the 5 percent total error, that is, 1.6 percent allowable error for each. The general case is that each component of a set of n components must have an error of no more than the total allowable error divided by the square root of the number of such components. This is meant as a guide. It would not be reasonable to assume all components have an equal error. If it is important, one should perform a more thorough error analysis.

The size, shape, and body posture of the cyclist affect the coefficient of drag. A change in body posture between the downhill and uphill portions of the test can change the outcome. Repeated measurement with different rider weights and clothing can be used to reduce errors.

7.1.2   Finding Coefficient of Drag

The suggested downhill/uphill testing provides the data for calculating the force of air resistance, which is one component of windage and friction (F_{wf}). However, one must remove the component of rolling resistance. Our measurement ensures that F_{wf}, air resistance and friction losses, are equal to the downhill component of gravity force, which we show above to be the product of bicycle and rider mass and road grade. A separate measurement following the procedure given in Chapter 2 of rolling resistance or the data thereof may suffice. Subtracting rolling resistance from F_{wf} determines air resistance. Knowing the force of air resistance we can deduce the product of frontal area and coefficient of drag; see Chapter 2. We determine cross section of the frontal area from photographs, either digital, video, or film, taken of the rider and bicycle head-on along with a standard of length. Enlarging such photographs will assist in obtaining an accurate determination of frontal area. Dividing the deduced product of frontal area and coefficient of drag by the frontal area determines the coefficient of drag.

7.2 MEASURING MOTOR EFFICIENCY INCLUDES MEASUREMENT OF MOTOR POWER

Although motor efficiency can be high, even greater than 90 percent, the wide range of travel conditions requires motor speeds that often go beyond the high-efficiency operating range of a motor. Since we cannot assume high efficiency, and thus ignore it in our bicycle operation, we must know what it is. Although motor manufacturers have these data, they may not be available. Typically, measurement of motor efficiency is a specialty field, and for best results it is best left to motor experts. We present here, an outline of measurement techniques that could be implemented by the nonexpert.

7.2.1 Motor Output Power and Efficiency

Measurement of motor efficiency requires measurement of motor input and output power. Efficiency is determined by the equation: $\eta = 100 \times P_{out}/P_{in}$, where η is expressed as a percentage, P_{out} is the output power, in watts, and P_{in} is input power in watts. Input power is the product of current and voltage at the motors electrical terminals. Output power is obtained by measuring speed and torque the motor delivers. However, sometimes this output power is converted to a voltage and current that represents the mechanical power. An example is a dynamometer that is a calibrated electrical generator that feeds a resistive load. The resistive load dissipates the power output of the motor being tested. Ammeters and voltmeters, or their equivalents, indicate the power flowing into the resistive load. By equivalents we mean digitally processed current and voltage that, when multiplied together, give a power value that accurately corresponds to the power delivered on the shaft of the motor being tested.

For the nonexpert we suggest the prony brake test method of measurement. This method uses friction on the motor shaft while the shaft exerts a measurable force. One means of accomplishing this is to use a static belt and a pulley mounted to the motor shaft. Each end of the static belt has a tension-measuring device between the belt ends and a stable object, such as the floor or test bench table. Some means of tension adjustment permits varying the mechanical load on the motor. For each adjustment of the two tension forces the motor speed is recorded as well as the corresponding motor input voltage and current values.

A means of cooling the pulley and belt will likely be necessary. Otherwise the equipment will need to cool naturally between tension settings. Use of a digital data acquisition system allows for rapid data gathering and may permit a full range of tension adjustments without use of cooling. Repetition of the test will increase accuracy of measurement.

Recall from Chapter 3 that power for rotating devices is $P = T\omega$. For power in watts, T is expressed in newtons, and ω expressed in radians/second. For the belt and pulley method, T is the difference between the two tension values of the static belt times the friction diameter of the pulley. Refer to the Appendix for conversion of force and speed values to mechanical power.

Another implementation of the prony brake method is to attach a drum with a brake band around it to the motor shaft. If the motor is a "wheel hub motor," then just an exterior brake band is needed. A torque arm is attached to the brake. The force on the torque arm is measured while the brake friction on the drum or hub is varied. The torque in this case is the force on the torque arm times the radius from the motor axis to the point of the force. Motor speed and input voltage and current are recorded at the same time as above, and motor output power is calculated. Motor input power is calculated in the manner described below. If desired, a similar brake can be applied to the drive wheel of the bicycle. With knowledge of motor output power and wheel output power, one can determine the efficiency of the wheel drive mechanism.

It is possible to eliminate the use of the torque arm by using commercial devices that mount around the motor shaft or couple the shaft to a load. These devices can provide digital signals for recording. Force cell monitors or strain gauges are available for measuring the force.

7.2.2 Motor Speed Sensing

Motor shaft speed can be measured by rotation sensing or a tachometer. Tachometers can be mechanical or electrical devices. Mechanical measurement uses a rotation counter and a timer or Biddle indicator. Electrical measurement is recommended because with it the motor speed can be more accurately correlated with the force measurement, the motor input voltage, and current measurements. A calibrated small measurement motor, coupled to the bicycle motor under test by means of fiber or rubber disk, provides an electrical signal that is proportional to the test motor speed.

Rotation sensing can be accomplished with optical or magnetic sensors. Optical methods use light-emitting diodes (LEDs) for illuminating a rotating disk and an optical transistor for sensing either interruptions of light transmission or reflection variations during rotation of a object. Magnetic sensors can be used to measure speed by providing signal output changes from rotating magnets. The magnets can be individual or continuous in magnetic rings that have alternating north and south poles along a circumference. The sensors can be based upon variable reluctance or semiconductor sensitivity to magnetic fields. Variable reluctance sensing uses pickup coils that produce changes in coil output voltage for changes in a rotating magnetic field. Semiconductor sensing using solid-state

Hall effect devices to produce output voltage changes for either a rotating magnetic field or interruptions in a static magnetic field produced by a rotating steel vane. There is also a Hall effect device with a small magnet within its case. It is then used to sense rotating cogs or gear teeth.

7.2.3 Motor Input Power

Measurement of motor input power should occur at the terminals of the motor. The means of accomplishing this measurement will depend upon whether motor design is dc or ac. An ac motor can be designed to run on a single phase or multiphase ac.

To avoid the complications of an ac or brushless dc motor measurements, one might want to include the inverter and controller loss with the motor loss by simply measuring the current and voltage coming from the battery. However, in so doing it would be best to use rms meters because the time-varying load may cause time-varying current and voltage at the battery terminals. Also there may be a small amount of energy back flowing into the battery from the inverter.

Voltage and current for an electric motor reach their peaks at slightly different times, and at times during a cycle their magneto motive forces cancel each other. The rms meters or equivalent means are required to obtain an accurate measure of motor input power. In rms meters the moving element responds in a positive upward motion to the torque produced by both the negative and positive portions of an alternating current. In an ordinary dc meter the motion caused by the negative and positive portions of the alternating current cancel one another and the meter reads zero.

A test of the voltmeter, whether digital or not, is to measure an ac sinewave voltage with and without a diode in series with one of the measurement leads. The meter measures true rms values if the voltage with the diode in the measurement circuit is 70 percent of the measurement without the diode [1].

Equivalent means of determining the rms power is with a wattmeter or digital data acquisition system as discussed in Section 7.3.1. Use of the wattmeter method is limited by the upper-frequency response of the meter. The predigital era wattmeters could respond to frequencies up to about 1 kHz. However, more recent designs can respond to frequencies used to drive modern multiphase motors [2, 3].

Measurement of motor input power to multiphase motors requires special consideration. For electric bicycle applications it can be assumed all phases of the motor are balanced. With this assumption, measurement of input power to one phase is adequate, and the total power is the number of phases times the power input to one phase.

Current sense resistors are a common method of providing a means of measuring current. The sense resistor has a small resistance, typically less than 1 ohm (Ω). Sensing the voltage drop across the resistor provides the information for calculating the current: voltage drop across the resistor divided by the resistance. The resistance must be small enough so that the expected current will not significantly heat the resistor. Nor must it be so high as to significantly reduce the voltage supplied to the motor. Conversely the resistance must be high enough to provide for an accurate measurement. An example is 10 A flowing through a 0.022-Ω sense resistor. The voltage drop across the resistor, the product of the current and resistance, will be 220 mV. The power dissipated in the resistor will be the product of the current squared and the resistance, 2.2 W in this example. A power rating for this sense resistor is 5 W. Manufacturers or their data sheets should be consulted for the application in regard to accuracy of the resistance and its change when heated. The rms value of current and voltage must be determined first before multiplying the two values.

7.3   MEASURING BATTERY CHARACTERISTICS

7.3.1   Data Gathering

Data can be gathered many ways now. The original method of handwriting the data information down on paper sheets, notebooks, or specially prepared forms is certainly available. For our measurements where we are concerned about obtaining a congruent set of data, we would need up to four people to record simultaneously what is occurring and inform a data logger of the values. This procedure is difficult and prone to errors.

The next step would be to use strip chart recorders. They can be expensive and may not be portable. Paperless data loggers may be satisfactory for obtaining congruent data. However, paperless data loggers have relatively low sample rates and may have limited solid-state memory. They are constructed from proprietary circuitry, making purchase cost and the cost of replacement parts relatively high.

Presently, there are data acquisition systems that are tailored for use of personal or laptop computers that use plug-in cards or port attachments. The computer-based data acquisition systems use a combination of computer hardware and software and signal conditioning equipment that gathers, stores, and processes data for analysis, display, and reporting.

Strip Chart Recorders   Strip chart recorders may be adequate for laboratory applications if they are already in the possession of the tester. However, they are generally bulky and heavy and not intended for portable use. They consume paper and pens that may not function. In addition the pens may only provide monochrome recording, making overlapping signals indistinguishable. They also require thorough familiarity with their unique configuration and operating procedure and are constructed from proprietary circuitry and mechanisms, making purchase cost and the cost of replacement parts relatively high.

Test signals feeding the various recording pens must be modified to conform to the voltage range of the galvanometer and the driving amplifier. The galvanometer is what causes a recorder pen to move in response to the test signal being recorded. Furthermore, the test signal frequency may have to be translated to be within the frequency range of the recorder. Digital test signals will need to be converted to appropriately scaled analog signals.

The data recorded on a strip chart can be difficult to reduce to an easily analyzable form. One difficulty is correlating the data with events. Quick scribbles on the chart paper may not be readable. The recorded data may not have enough dynamic range. For example, it may be easy to make out volts but not small values of tenths of a volt on the recording paper. The recorder paper speed may be adequate for some test signals but not others during a test.

Digital Data Digital data acquisition was implemented years ago on dedicated hardware that cost thousands of dollars. Little or no digital data acquisition capability was available on personal computers (PCs) and less expensive systems. However, many vendors are now selling hardware boards and software packages that implement very powerful digital data acquisition capability.

Data acquisition is a very common and necessary tool in today's technology. The data acquisition system captures a series of samples of a signal at definite time intervals. Each sample contains information about the signal at that specific time. By knowing the exact time of each conversion and the value of the sample, one can reconstruct the sampled signal. This is true to the extent that the signal is sampled often enough. Sampling should occur at least at twice the frequency of the signal, but the signal can be more faithfully reproduced if sampled at five times its frequency. The characteristics of the unit under test determine the sampling frequency that should be used. The slowest signals are likely to be those associated with temperatures. During travel, faster signals such as speed and road surface need to be recorded. Electronic components on the bicycle are likely to produce the highest frequency signals. In particular the motor controller electronics, if motor drive waveforms need to be monitored, will be the source of the highest test frequencies. These waveforms with voltage changes occurring within microseconds will require sampling frequencies in the order of megasamples per second. This, however, is an area for the circuit specialist.

For torque measurements it is easy to see that one might want to know how the motor torque varies within a single rotation. The sampling rate is

$$s = 6k\,R_{\mathrm{pm}}/\alpha \quad \text{(samples/second)} \tag{7.1}$$

where k is a constant representing how faithfully a waveform is to be reproduced (2.5 to 5 suggested), R_{pm} is the motor rotation rate, revolutions/minute, and α is the rotation change between samples, degrees/sample. For travel one can use the relationship:

$$s = k\beta v_g/3.6 \quad \text{(samples/second)} \tag{7.2}$$

where k is as above, β is the number of samples/meter of travel, and v_g is the travel speed, kilometers/hour. An example would be for travel at 20 km/h, with the need to get one sample for each wheelbase length of bicycle travel. Assuming the wheelbase length is 1.1 m, then we find that $\beta = 1/11 = 0.91$ and $k = 1$ in this instance. Then $s = 5.05$ samples per second.

The number of binary bits used in sampling defines the incremental steps to which the signal voltage can be reported. The computer uses these digital values to recreate the signal waveform. The increments of signal voltage, δV, are

$$\delta V = V/2^b \tag{7.3}$$

where V is the signal voltage range, and b is the number of binary bits. For 8 bits that would be $V/256$ volt increments and for 12 bits it would be $V/4096$ volt increments.

The test signals at their source in the equipment are likely to not be well suited for direct connection to the analog-to-digital (A/D) converter used for data acquisition. These signals may need to be amplified, attenuated, or modified in some other way. For example, assume the A/D has a functional range of 0 to 5 V. If the test signal has a range of -10 to $+10$ V it will be necessary to attenuate it by a factor of 4 and offset the most negative value to 0. It is best to condition each test signal so that its range equals or is slightly less than the A/D range.

A data acquisition system for use with a computer includes an interface unit for interfacing with a computer, a control data memory for storing control data that is associated with controls for the computer; and a file data memory for storing file data that is to be acquired by the computer. A processing system is provided for controlling transfer of data to the computer memory system, which has hard drives or flash memories, depending upon the computer used.

A large majority of digital data acquisition systems available on the market for PCs and similar workstations, as well as laptops, have their own software package for allowing a vendor's hardware (plug-in boards, cables, sensors, etc.) to be controlled by the computer. The advantage of each vendor providing its own software is that the vendor then has control over how the hardware is accessed, controlled, and used, thus ensuring correct operation of the entire digital data acquisition system. The vendor's software can be a good interface to other digital data acquisition programs if used, such as plotting, graphing, and data recorder programs. However, the data acquisition software itself is likely to include plotting and graphing ability.

One disadvantage of each vendor providing its own digital data acquisition interface software is interoperability. If a user likes the plotting utility provided by vendor X's digital data acquisition package, but needs the performance characteristics of vendor Y's plug-in board, it is almost guaranteed that the two pieces will not play well together. In today's torrid environment of new, faster computers coming out almost monthly, interoperability of digital data acquisition hardware and software between different systems would be highly desirable.

Two other disadvantages of each vendor providing its own digital data acquisition interface software are cost and learning curves. Typically, each vendor provides a software package that is either proprietary at worst, or a pseudo-industry standard at best. One has to go to this vendor for other software such as analysis packages and signal processing packages that work with the digital data acquisition hardware and interface software, and so forth. This holds one into a single vendor to accept a less-than-preferred software package as a result, just due to its software interface. With each new vendor's digital data acquisition interface software, the user must then learn how to use the package, work through any bugs, and get proficient with it to the point where the digital data acquisition work becomes efficient and productive.

Martin [4] shows us how a digital data acquisition system can be provided to mimic a hard drive. It takes advantage of the fact that application software is insulated from low-level hardware by the operating system to achieve not only application software independence but also system portability and interoperability. One area of commonality on almost all computers that perform digital data is hard disk storage. Hard drives are a fundamental part of almost all computers, and users are generally comfortable with using them and opening and closing files, copying and deleting files, organizing their disk, and the like.

The disadvantages discussed may not apply to particular requirements. Discussions with vendors in the business of data acquisition and test data analysis are a requirement prior to test design.

REFERENCES

1. Herman P. Raab, True RMS Operation Test, *Supplement to Electronic Design*, October 23, 1997.
2. W. Stephen Woodward, Simple Digital AC Wattmeter, *Supplement to Electronic Design*, October 22, 1998.
3. Robert A. Pease, What's All This Wattmeter Stuff, Anyhow?, *Electronic Design*, May 13, 2002.
4. Thomas J. Martin, Data Acquisition System, U.S. Patent 6,105,016, August 15, 2000.

CHAPTER 8

DEVELOPMENTS TO WATCH

Battery-powered electric bicycles will become important means of personal movement throughout the world as the cost of alternative means grows. This need will accelerate technical developments, which are already occurring at ever-increasing speed. For example, the growing market for lightweight laptop computers motivated in 5 years a new battery industry that in 1997 produced rechargeable lithium batteries worth $10 billion. Developments that will affect electric bicycle performance and cost include:

- Large-cell lithium batteries
- Zinc–air batteries
- Nickel–metal hydride batteries
- Fuel cell sources of electric power
- High-efficiency coreless motors
- Solar charging systems
- Further application of microcontrollers and microprocessors
- Molded carbon-fiber-reinforced bicycle frames
- Comfortable rain-resistant cold-weather clothing
- World's population growth

New developments are going to come rapidly. Three areas in which important developments need to be watched are:

1. The world's petroleum production is going to decline, as noted in Chapter 3.

2. Nations that encourage electric bicycle use are economically producing exportable products that compete very effectively in world markets. For example, China's electric bicycle sales were over 4 million in the year 2003.

3. New developments in energy storage, propulsion motors, and control are reducing the cost of bicycle travel and increasing the travel distance available to riders of electric bicycles. A rider who goes 100 miles on an electric bicycle would have to pay 8 cents for the energy if it cost $0.13/kWh. Gasoline at $2.50/gal, for traveling 100 miles in an SUV, would have cost $9.00 in the year 2004.

Electric Bicycles: A Guide to Design and Use, by William C. Morchin and Henry Oman
Copyright © 2006 The Institute of Electrical and Electronics Engineers, Inc.

8.1   BICYCLE SYSTEMS

8.1.1   Frame Design, Clothing, and Energy Management

Molded Carbon-Fiber-Reinforced Bicycle Frames   Bicycle frames are traditionally built by welding or brazing high-strength tubing into fittings like the bottom bracket and headset. This construction has produced sturdy bicycles that have been proven to be safe in decades of service. New materials technology is changing the designs of other manufactured products. For example, automobile bodies are made by welding together hundreds of metal pieces. Electric car builders find that they can mold the body with durable plastics for a fraction of the cost of traditional bodybuilding. Airplane manufacturers have found that with carbon fiber reinforcement even flight-safety-critical structures can be made with plastics rather than metal.

Carbon-fiber-reinforced bicycle frames are already being built. Feasible options include provisions for mounting propulsion motors and enclosed space for batteries in structures that also carry stresses. Electric wiring could be enclosed in the frame structure to protect it from moisture and accident damage. Frame members could also be designed to minimize air drag.

Today's bicycle configuration has evolved from the first bicycles. Human factors studies have changed many traditional body positions to achieve efficient results. For example, the human-powered Raven airplane is designed for a 5-h 100-mile nonstop flight. Modeling the airplane on a computer, which could calculate its performance, minimized the required propulsion power.

Configuration variations were tested in the model, and the lowest power design was adopted. However, this design was dynamically unstable, and it also approached the power output limit of an athletically conditioned pilot. The pilot would be overstressed if he had to continuously control the airplane's attitude, as well as pedal vigorously. Adding a 1-lb battery-powered autopilot that moved the rudder to damp out the instability solved this problem.

Applying today's human factors knowledge into the electric bicycle design could produce a surprising comfortable mode of personal transportation.

Comfortable Rain-Resistant Cold-Weather Clothing   Bill Woods bicycled over interesting routes in the Northwest United States and wrote books describing them in pertinent detail. His comment on clothing was: "You are going to get wet from sweat or rain, so wear the minimum clothes." The heat from hard pedaling could be absent when riding an electric bicycle. Also, today's clothing technology has provided comfort for tasks ranging from deep underocean diving to working in a spacecraft.

Now-available fabric and insulation technology makes possible the design and fabrication of clothing that can be worn over work clothes to keep the bicycle rider comfortable. The electric bicycle rider does not need to have freedom of continuous movement of his legs. He does not need to stand up to pump up steep hills. Furthermore, he would appreciate not having to take a shower after he arrives at his workplace on a bicycle. Meeting these challenges is within the capability of the clothing industry.

Energy Management of Power Sources New application of algorithms in microprocessors for minimizing energy expenditures in bicycle propulsion is a possibility. Inclinometers within the circuitry and operator input of desired travel parameters and digital map data can be used by a microprocessor to compute and assist in controlling the bicycle speed for minimum energy to meet specified travel parameters.

8.1.2 Bicycle Transportation Systems

World's Population Growth The world's population doubled between 1830 and 1930. By 1987 it was growing at a rate of 488,000 per year. Faster growth can be expected in the future. Transportation of people will become a major challenge in the twenty-first century. Constructing new freeways on which people drive to work is becoming difficult. Even widening existing freeways to increase their capacity is impossible in many cities. Time-consuming traffic jams result. Saint Petersburg, Russia, offers an example of how to handle population growth. Constructing big downtown apartments is no longer practical, so they built a village of 20-story apartments 10 miles beyond the city's borders. A passenger railroad connects the village to downtown Saint Petersburg, so very little parking space for cars is provided in the village. However, in addition to traveling downtown to work, the people would like to tour around the countryside and visit the beach. Electric bicycles could supply this transportation need, as well as the travel from the downtown station to the workplace.

An important development to watch is the possible effect on the U.S. trade deficit as bicycles replace automobiles for personal travel in other nations. In the United States our trade deficit will grow and the value of our dollar will decline because we will have to import even more petroleum as our oil resources approach exhaustion. These developments can create in this nation a growing workforce of workers that are employed in low-paying jobs such as retail sales, delivery services, and maintenance of yards, homes, factories, and stores. In China, where low-income jobs predominate, bicycles are in common use. Use of electric bicycles is growing, and production of electric bicycles was over 2.5 million a year in 2004, nearly a 35 percent annual increase from year 2002.

Private and Public Partnerships An advantage of an electric bicycle is that it does not need a $20,000 parking spot. An electronic-valet parking lot could store electric bicycles in an apartment complex and also at the downtown train station. Furthermore, a standard automobile parking space can hold eight bicycle parking spaces

Building roads and maintaining them is expensive, costing millions of dollars per lane mile. Bikeways, with well-maintained shoulders and sidewalks are far less expensive. In 1994, a typical bikeway cost from $250,000 to $500,000 per mile. It can accommodate as much or more traffic than a freeway lane [1]. And in dense urban traffic, the cyclist usually gets around town faster than automobiles. In the United States, where the bicycle is used the least of all and where the greatest gain can be made in conservation of energy, there needs to be the most cooperation between the public and private sectors. The private sector needs to take the initiative, but the public sector can form the leadership. The partnership can be one where the citizenry (the private sector) performs the studies and inventories of what needs to be done to accommodate and encourage the cyclist. The public sector needs to guide the private sector in financial, engineering, and regulatory matters. Public planners need to be educated and convinced to change their views of automobile and public transportation.

City Transportation Systems   Problems exist in the United States that restrict bicycle ridership to a hardy few. For the rider, these include danger from autos, stop-and-go driving on streets not designed and maintained for skinny tires, and bad weather. Additionally, bicycle traffic slows auto traffic when the two are mixed. And for the local government, bicycle lanes take up valuable space without providing revenue. Brown [2] has presented a concept of an elevated, enclosed bicycle transportation system that would enable the electric bicycle to become an acceptable form of transportation for the general public and reduce air pollution and dependence on foreign oil. It would be cantilevered above existing city streets. Jamerson [3] presents a similar system, called Trans Guide 2000, that was considered in Norway.

Each concept uses an elevated enclosed bikeway. Figure 8.1 shows Brown's concept. Although the figure shows access from a nearby building, the concept is not limited to that type of access alone. Access to the bikeway could be provided with ramps, elevators and other means. Brown suggests an air-conditioned enclosure, and the Trans Guide concept uses induced wind in the direction of bicycle travel. The idea is to have a wind speed at the backs of bicyclists to counter the effect of air resistance. The elevated bikeway would separate two-way travel so

Figure 8.1   Elevated covered bicycle roads [1].

that those either going or coming would have wind to their backs. The claim is that bicyclists will be able to ride 10 km inside the enclosure for the same amount of energy they would use to ride 1 km outside the system.

A complete bicycle road system would consist of a grid of roads above city streets, spaced a few blocks apart in the Brown idea. Employers could have entrances and storage lockers at the bicycle road level. Bikeways from the ends of the city grid would fan out into the suburbs about 10 km from the city center. Cities with subway systems could have bicycle roads from the subway stations to nearby housing communities, reducing the parking problem at stations. Brown's cost estimate is that the bikeway would equal the cost per unit distance of a two-lane asphalt road in a rural area.

Brown estimates the bikeway can permit about 3200 people per hour to move in each direction if spaced 10 m and traveling at 25 km/h in a two-way corridor of 4.5 m width.

Highway and Freeway Travel   Existing highways and freeways are a wasteland for bicycles. Planners, transportation designers, and governments must come to realize that traffic congestion can be much relieved if bicycles are permitted use of these roads. With proper safeguards such use should prove possible. For a bicyclist, direct routes and faster travel between destinations would be possible.

The degree of protection of the bicyclist can be varied according to the density and speed of automobile traffic that would share the highway or freeway. In-city routes can use elevated bike ways similar to the Brown design. Less congested locations such as suburban areas could use fenced elevated surfaces for bicycles. In rural areas low fence or concrete bulkhead separators could be used.

Coming Events: Discoveries in Scientific Research   New technology is enabling scientists to discover how nature's efficient processes work. Pertinent to vehicles is the high efficiency of muscles that people and animals use to deliver propulsion power. Muscles are composed of very tiny fibers that ratchet over each other to create a tension when electrically commanded to do so. The process consumes sucrose and oxygen. It enables a dolphin to swim 3000 miles with food containing the energy equivalent of 1 gal of gasoline. We have not yet learned how to utilize the muscle power process in propulsion engines. Being able to extract pure oxygen from the air would make possible development of new high-efficiency bicycle propulsion engines.

8.2 ENERGY SOURCES

8.2.1 Batteries

Large-Cell Lithium Batteries   Lithium cells in sizes appropriate for electric bicycles are being built for prototype electric cars. The materials used in making the cells are relatively inexpensive, and many automated manufacturing processes have been developed. The lithium battery will be a prime candidate for powering electric bicycles as soon as factories start producing 20- and 30-Ah cells. For electric bicycles the energy content of the battery is very important. A lithium battery, which delivers 120 Wh/kg, will carry the rider 5 times the distance that he could get from a lead–acid battery that delivers 24 Wh/kg.

However, proving the charge–discharge cycle life of a battery in its expected operating environment requires time. The batteries must be tested in chambers that can maintain temperatures ranging from the subzero of northern winters to the over-100°F that occurs in deserts in the summer. A battery that survives 1000 charge–discharge cycles would be attractive for powering a bicycle.

We already know that lithium batteries will require a "smart" charger. However, to simplify the charging process, electrochemical solutions to the charge-unbalance problem are also being developed.

Nickel–Metal Hydride Batteries   The cost of the materials for a nickel–metal hydride battery can become significant for battery sizes appropriate for powering an electric bicycle. This battery is similar to a fuel cell in that during discharge hydrogen combines with oxygen to produce water. During charge the hydrogen is recovered by electrolyzing this water. The hydrogen gas is stored in the metal hydride. The metal hydride consists of expensive compounds, such as titanium, zirconium, vanadium, manganese, and palladium [4]. The required rare-earth elements are by-products from mines and refineries that produce other marketable products. A big demand for these elements could cause the price to rise. On the other hand, researchers are testing other hydrogen storage compounds. Also, charge–discharge cycle life tests are in progress. The results of these developments will determine if the nickel–metal hydride battery is the best choice for powering electric bicycles.

Zinc–Air Batteries   A zinc–air battery cannot be recharged with a simple charger. The cassettes must therefore go into a recharging station where automatic machines remove the zinc oxide, electrochemically recover the zinc, and assemble new cassettes. Without this infrastructure zinc–air batteries are not practical. However, the infrastructure need not be local. For example, in desert sunshine a 10-m by 10-m solar cell panel can one day generate the energy required to recharge the cassettes from bicycles that have traveled 9160 miles. In a developed infrastructure, the recharged cassettes could be available at a local store, and the bicycle owner would have a supply at his home. Discharged cassettes would be shipped to a recharging station near a solar, nuclear, or possibly a future fuel cell power plant. Charge–discharge cycle life would no longer be an issue because the recharged cassettes would be indistinguishable from new ones.

Other Batteries   Most of the newly invented batteries show attention-getting performance but also have basic flaws that are discovered later. For example, as discussed in Chapter 3, the rechargeable nickel–zinc cell appeared to be low in cost, but zinc whiskers grew on the negative plates each time the cell was recharged.

8.2.2   Fuel Cells

In a common form of fuel cell hydrogen gas is combined with oxygen to produce electric power. Fuel cell power plants have been built, but they have been abandoned because hydrogen made from the methane in natural gas carries trace impurities. These impurities damage the fuel cell plates, adding maintenance expense.

Zinc–Air Fuel Cell — A Development to Watch   In electric bicycle applications the cost of the energy consumed in powering the bicycle can affect its usefulness. For example, the cost of commuting to work by poverty level persons in China and India is very important. A long travel range for each energy replenishment is also important in all travels because a source of energy might not be conveniently available at a location where the bicycle's energy supply is exhausted.

Zinc–air fuel cells are now delivering 400 Wh/kg. Propelling a bicycle over a typical road consumes around 15 Wh/mile (10 Wh/km), depending on the bicycle's travel speed. Thus 1 kg of zinc fuel cell pellets will propel an electric bicycle a distance of about 25 miles (40 km). This performance makes the zinc–air fuel cell a development to watch.

8.2.3   Ultracapacitors and Fuel Cells to Replace Batteries

Chapter 6 indicated experimental ultracapacitors have energy densities of 27 to 60 Wh/kg. Such capacitors have operating voltages of about 3.8 V. The 27-Wh/kg rated capacitor provides a specific power of 8 kW/kg, and the 60-Wh/kg capacitor has a specific power of 540 W/kg. The specific energy values are comparable to the lead–acid, nickel–metal hydride, and the nickel–cadmium batteries. A series connection of 6 to 9 such capacitors would supply the voltage and power for the electric bicycle. However, it may be possible to use one capacitor if one uses a charge pump technique that multiplies the capacitor voltage, following the teaching of Nebrigic et al. [5]. Otherwise, a series connection of capacitors will require the cell equalization techniques similar to those used for lithium battery cells described in Chapter 4.

An ultracapacitor combined with a fuel cell can be used to enhance the travel range by providing for increased motor efficiency during periods of uphill travel. Furthermore, Dowgiallo and Hardin [6] say that greater than 100,000 cycles of life are expected for the ultracapacitor.

8.3   SOLAR CHARGING SYSTEMS

8.3.1   Storing Solar Energy in Zinc — A Development to Watch

Free energy from solar power could be an environmentally desirable source of bicycle propulsion power for some bicyclists in the United States. It could be a practical energy source for expanding educational opportunities in the jungle region of Africa. However, the cost of constructing roads, plus pipelines for bringing in petroleum fuels, is beyond the presently available resources of these communities. Bicycle paths could be constructed with local labor, and even over steep hills, so that the students could bicycle to college every morning. Electric bicycles could double the travel distance and quadruple the area served by the college.

Solar energy, which is not available at night or on cloudy days, can be stored in batteries or hydrogen tanks. The hydrogen tanks would be huge unless the gas is stored at high pressure produced by a compressor. A battery is a costly electrochemical tank that contains many sophisticated components. Heavy lead–acid batteries that are often used in energy storage will degrade in capacity as they age. On the other hand, zinc pellets can be stored in pertinent quantities at low cost in plastic-lined boxes.

For supporting a tropical-area community college, the key components could be the zinc–air fuel cell and the zinc–air "filling station" shown in Figure 3.17. The filling stations would be located at the community college. At the stations every student would deliver during each school day the zincate fluid from the zinc–air fuel cell on his or her electric bicycle. Then the student would receive enough potassium hydroxide electrolyte and zinc pellets for bicycling home and returning to school the next morning.

8.3.2   Thermovoltaic Cells

A thermoelectric generator [7], which converts heat directly from a heat source into electrical energy, can be applied to charging batteries. The devices, however, only provide useful electricity at elevated temperatures, typically 1200°C. The output is about 0.8 W/cm^2 of cell area at a voltage of 0.38 V. The cells have been combined with photovoltaic cells for space applications. In addition thermovoltaic panels have been combined with wood-burning stoves to provide 100 W of power at 12 V dc [8].

The cells create electricity with the application of an external electromotive force (emf) across a heated semiconductor to produce the drifting of electrons and thereby producing a current in the semiconductor that can be used in a load. Operation is possible at maximum efficiency near 80 percent. When used in combination with a concentrating lens or combination of a heat absorber and heat exchanger, the cells can convert the sun's energy to electricity [7].

8.3.3   Integration of Flexible Solar Cells with Batteries

Solar cell technology is advancing to the point where flexible plastics have semiconductor material deposited on them. A thin-film flexible solar cell is built in this manner. A cadmium telluride p-type layer and a cadmium sulfide n-type layer are sputter deposited onto the plastic at a temperature sufficiently low to avoid damaging or melting the plastic. A transparent conductive oxide layer overlaid by a bus bar network is deposited over the n-type layer. A back contact layer of conductive metal is deposited underneath the p-type layer that completes the current collection circuit. The semiconductor layers may be amorphous or polycrystalline in structure.

Solar panels when partially shadowed show a significant loss in output power. Hence a cylindrical or other curved panel surface that encounters shadowing will not provide power as expected. However, segmented sections that are not individually shadowed will provide nominal power. It may be possible to create an aerodynamic surface with a solar panel system to both produce power and reduce drag. Parking the bicycle with the proper orientation to the sun could provide some degree of battery charging while the cyclist is otherwise occupied. Roughly, in a good Arizona sunshine, one could realize about 13 Wh/m^2/day of solar panel energy with a planar panel.

8.4   HIGH-EFFICIENCY MOTORS

Permanent magnet synchronous motors are now delivering high efficiencies, on the order of 90 percent, that were never before attainable in synchronous motors that had energy-consuming field coils, nor in induction motors that had energy-consuming rotor bars. Long riding ranges on electric bicycles become available when very little battery energy is wasted in motor losses.

High-speed lightweight motors need to be coupled to the bicycle's pedal shaft or propulsion wheel with speed reduction gearing that has power losses. A motor that is directly coupled to a wheel would rotate at a slow speed, and would be heavy, but losses in speed reduction gear would be eliminated. A successful alternative is the hub motor, which is built into an enlarged hub of the rear wheel. The motor diameter and gear ratio are optimized to achieve a combination that gives the greatest travel distance per watt-hour of energy consumed from the bicycle's battery.

A synchronous motor could have a stator that is attached to a bicycle's fork. The stator winding would be configured to pull on passing field magnets that are mounted on the side of the rim of a bicycle's wheel. This motor could deliver propulsion power with the highest possible efficiency because losses in speed reduction gearing are avoided. However, many costly permanent magnets would have to be purchased and mounted on the rim. An alternative squirrel cage rim motor would have a "rotor" that is a combination of iron pads and aluminum bars on the side of the wheel rim. Some of these motors have been built (see Fig. 5.6), but their performance has not yet been reported in data that can be analyzed and optimized. This is a development that needs to be watched.

A possible obstacle to efficient bicycle motors is the cost of rare-earth magnets. The magnets for a prototype motor would cost around $100 in 2005 dollars. Quantity prices would be lower. However, permanent magnet motors would also enable automobiles to travel farther on a battery charge. The automobile market could consume so much rare-earth elements that the price would rise until new mines are opened to produce these elements. Today they are a by-product of precious metal mines.

8.5   CONTROLLERS

Electric motor speed control technology has been developed for industrial and propulsion applications. Thoroughly documented in design manuals and computer programs is the procedure for selecting the speed control configuration, and its components, for a given motor classification that fulfills the requirements of the machine being driven. Availability of this data resulted from the need of precisely controlled motor speed for many industrial applications. Also, sophisticated computer chips are available for performing motor control functions such as are required for electric and hybrid electric cars.

Bicycle applications vary from simple shopping trips in the rider's neighborhood to vacation trips that cross mountains. Presently, available bicycle motor controllers will make the bicycle travel at a speed that the rider selects but slows the bicycle's speed if climbing a steep hill if the selected speed would overload the motor.

A motor control that would also detect overstress on a lithium ion battery that supplies propulsion power is also possible. A lithium battery is a good energy source because it can deliver over 133 Wh/kg of mass. However, completely discharging a lithium ion battery could damage some of its cells if they are at different temperatures. Continuous reporting of the state of charge the bicycle's battery to the bicycle rider would be a useful service. Sophisticated motor controllers are being developed for electric cars. These controllers are developments to watch because they can be adaptable to electric bicycles.

Controllers should become smaller with advances in the semiconductor industry. Half-bridge motor drivers capable of driving motors up to 250 W without using heat sinks are now available on a single semiconductor chip, and up to 400 W if it is heat sinked. Three such units, necessary to drive a three-phase dc brushless motor, may not be far behind. The microprocessors to control these drivers are also on a single chip. We may see such designs evolve to a controller contained on a few chips mounted in one semiconductor package. Integration within the motor, now occurring, will become more prevalent.

REFERENCES

1. Michael D. Skehan, Integrating Rail Transit and Bicycles in Kent, WA, American Association for the Advancement of Science, 75th Annual Meeting of the Pacific Division, June 22, 1994.
2. Michael A. Brown, Electric Bicycle Transportation System, IECEC 2002 Paper No. 20112.
3. Frank E. Jamerson, *Electric Bikes Worldwide 2002: With Electric Scooters & Neighborhood EVs*, January 15, 2002, 6th ed., Electric Battery Bicycle Company. Frank Jamerson, Publisher, Naples, FL. can be contacted at e-mail elecbike@aol.com or www.EBWR.com <http://www.EBWR.com>.
4. Stanford R. Ovshinsky and Rosa Young, High Power Nickel-Metal Hydride Batteries and High Power Alloys/Electrodes for Use Therein, U.S. Patent 6,413,670, July 2, 2002.
5. Dragan Danilo Nebrigic, Milan Marcel Jevtitch, Vladimir Gartstein, William Thomas Milam, James Vig Sherrill, Nicholas Busko, and Peter Hansen, Ultra-capacitor Based Dynamically Regulated Charge Pump Power Converter, U.S. Patent 6,370,046, April 9, 2002.
6. Edward J. Dowgiallo and Jasper E. Hardin, Perspective on Ultra Capacitors for Electric Vehicles, *IEEE Systems Magazine*, August 1995, pages 26–31
7. James N. Constant; Thermoelectric Generator, U.S. Patent 4,292,579, September 29, 1981.
8. JXCrystals Internet website, www.jxcrystals.com, April 9, 2004.

APPENDIX

Table of Conversion Factors for Units of Measure

Symbol	Given $\Rightarrow$ Mult. by To Obtain $\Leftarrow$ Divide by	to Obtain Given	Symbol
FORCE			
oz	ounces-force	0.278 newtons	N
lb	pounds-force	4.448 newtons	N
kg	kilograms-force	9.807 newtons	N
kp	kilopond	9.807 newtons	N
dyn	dynes	1×10^{-5} newtons	N
oz	ounces-force	1/16 pounds-force	lb
kg	kilograms-force	2.2046 pounds-force	lb
kp	kilopond	2.2046 pounds-force	lb
dyn	dynes	2.248×10^{-6} pounds-force	lb
oz	ounces-force	0.0283495 kilograms-force	kg
lb	pounds-force	0.45359 kilograms-force	kg
kp	kilopond	1.02×10^{-6} kilograms-force	kg
dyn	dynes	1.02×10^{-6} kilograms-force	kg
DISTANCE			
in	inches	2.54 centimeters	cm
ft	feet	30.48 centimeters	cm
in	inches	0.0254 meters	m
ft	feet	0.3048 meters	m
m	meters	100 centimeters	cm
mi	statute miles	1.60934 kilometers	km
nmi	nautical miles	1.852 kilometers	km
AREA			
cmil	circular mils	0.0005067 sq. millimeters	mm^2
in^2	sq. inches	645.16 sq. millimeters	mm^2
cm^2	sq. centimeters	100 sq. millimeters	mm^2
m^2	sq. meters	1×10^6 sq. millimeters	mm^2
cmil	circular mils	5.067×10^{-6} sq. centimeters	cm^2
in^2	sq. inches	6.452 sq. centimeters	cm^2
m^2	sq. meters	10000 sq. centimeters	cm^2
ft^2	sq. feet	929.03 sq. centimeters	cm^2

(*continued*)

Electric Bicycles: A Guide to Design and Use, by William C. Morchin and Henry Oman
Copyright © 2006 The Institute of Electrical and Electronics Engineers, Inc.

Symbol	Given $\Rightarrow$ Mult. by To Obtain $\Leftarrow$ Divide by		to Obtain Given	Symbol
cmil	circular mils	5.0684×10^{-10}	sq. meters	m^2
in^2	sq. inches	6.4516×10^{-4}	sq. meters	m^2
ft^2	sq. feet	0.092903	sq. meters	m^2

VOLUME

fl-oz	fluid ounce (US)	29.57	cubic centimeter	cm^3
in^3	cubic inches	16.387	cubic centimeter	cm^3
ft^3	cubic feet	28316.85	cubic centimeter	cm^3
m^3	cubic meters	1×10^6	cubic centimeter	cm^3
l	liters	1000	cubic centimeter	cm^3
fl-oz	fluid ounce (US)	2.9574×10^{-5}	cubic meters	m^3
in^3	cubic inches	1.6387×10^{-5}	cubic meters	m^3
ft^3	cubic feet	0.0283168	cubic meters	m^3
l	liters	0.001	cubic meters	m^3
fl-oz	fluid ounce (US)	2.957×10^{-4}	liter	l
in^3	cubic inches	0.0163871	liter	l
ft^3	cubic feet	28.31685	liter	l
gal	gallons (US)	3.785	liter	l
gal	gallons (CAN)	4.546	liter	l

SPEED

mph	miles per hour	0.44704	meters per second	m/s
km/h	kilometers per hour	0.2777777	meters per second	m/s
kn	knots	0.5144	meters per second	m/s
mph	miles per hour	1.609344	kilometers per hour	km/h
kn	knots	1.852	kilometers per hour	km/h

MASS = weight/acceleration of gravity

oz	ounce	28.34952	grams	g
lbs	pounds	453.5924	grams	g
oz	ounce	0.0283495	kilograms	kg
lbs	pounds	0.4535924	kilograms	kg

DENSITY

lb/ft^3	pounds per cu foot	16.02	kg per cu meter	kg/m^3
lb/in^3	pounds per cu inch	27680	kg per cu meter	kg/m^3
g/cm^3	grams per cu centimeter	1000	kg per cu meter	kg/m^3
lb/ft^3	pounds per cu foot	0.01602	grams per cu centimeter	g/cm^3
lb/in^3	pounds per cu inch	27.68	grams per cu centimeter	g/cm^3
lb/gal	pounds per gallon	0.1198	grams per cu centimeter	g/cm^3

POWER

ft-lb/s	foot-pounds per second	1.355818	Watt (Newton-meter/sec)	W (Nm/s)
ft-lb/min	foot-pounds per minute	0.0226	Watt (Newton-meter/sec)	W (Nm/s)

Symbol	Given ⇒ Mult. by To Obtain ⇐ Divide by		to Obtain Given	Symbol
in-lb/s	inch-pounds per second	0.113	Watt (Newton-meter/sec)	W (Nm/s)
kpm/s	kilopond-meter/sec	9.807	Watt (Newton-meter/sec)	W (Nm/s)
oz-in/s	ounce-inch/sec	7.06×10^{-3}	Watt (Newton-meter/sec)	W (Nm/s)
oz-in/min	ounce-inch/minute	1.17×10^{-4}	Watt (Newton-meter/sec)	W (Nm/s)
BTU/min	British thermal units	17.58	Watt (Newton-meter/sec)	W (Nm/s)
BTU/hr		0.2930711	Watt (Newton-meter/sec)	W (Nm/s)
hp	horsepower(elec.)	746	Watt (Newton-meter/sec)	W (Nm/s)
kc/s	kilocalories/sec	4184	Watt (Newton-meter/sec)	W (Nm/s)

TORQUE

Symbol		Mult. by	to Obtain Given	Symbol
J	joules	1	Newton-meter	Nm
in-lb	inch-pound	0.113	Newton-meter	Nm
ft-lb	foot-pound	1.356	Newton-meter	Nm
oz-in	ounce-inches	7.06×10^{-3}	Newton-meter	Nm
kg-m	kilogram-meters	9.807	Newton-meter	Nm
g-cm	gram-centimeters	9.807×10^{-3}	Newton-meter	Nm
kpm	kilopond-meter	9.807	Newton-meter	Nm
ft-lb	foot-pound	0.1383	kilogram-meters	kg-m
oz-in	ounce-inches	7.2031×10^{-4}	kilogram-meters	kg-m
oz-in	ounce-inches	0.01388	gram-centimeters	gm-cm
g-cm	gram-centimeters	1×10^{-5}	kilogram-meters	kg-m
kpm	kilopond-meter	1	kilogram-meters	
oz-in	ounce-inches	5.2083×10^{-3}	foot-pound	ft-lb
g-cm	gram-centimeters	7.233×10^{-5}	foot-pound	ft-lb

MOMENT OF INERTIA

Symbol		Mult. by	to Obtain Given	Symbol
oz-in^2		182.9	gram-centimeters sq.	gcm^2
lb-in^2		2930	gram-centimeters sq.	gcm^2
lb-ft^2		421434	gram-centimeters sq.	gcm^2
kgm^2		1×10^7	gram-centimeters sq.	gcm^2
oz-in^2		1.83×10^{-5}	kilogram-meter sq.	kgm^2
lb-in^2		2.93×10^{-4}	kilogram-meter sq.	kgm^2
lb-ft^2		0.0421434	kilogram-meter sq.	kgm^2

ANGULAR VELOCITY

Symbol		Mult. by	to Obtain Given	Symbol
rpm	revolutions/minute	0.10472	radians/sec	r/s
rpm	revolutions/minute	6.28318	radians/min	r/m
°/s	degrees/sec	0.6283184	radians/sec	°/s

ANGULAR ACCELERATION

Symbol		Mult. by	to Obtain Given	Symbol
r/min^2	rev/min/min	1.745×10^{-3}	radians/sec^2	r/s^2
r/s^2	rev/sec/sec	6.283	radians/sec^2	r/s^2

LINEAR ACCELERATION

Symbol		Mult. by	to Obtain Given	Symbol
ft/s^2	feet/sec/sec	0.3048	meters/sec/sec	m/s^2
ft/s^2	feet/sec/sec	0.6818	mph/sec	miles/h/s

(*continued*)

Symbol	Given $\Rightarrow$ Mult. by To Obtain $\Leftarrow$ Divide by		to Obtain Given	Symbol
mph/hr/s	miles/hour/sec	0.447	meters/sec/sec	m/s^2
km/hr/s	kilometers/hr/sec	0.2778	meters/sec/sec	m/s^2
WORK and ENERGY				
	BTU (TC)	1054.35	joules (watt-sec)	J
	BTU (IT)	1055.056	joules (watt-sec)	J
	BTU (mean)	1055.87	joules (watt-sec)	J
	foot-pounds	1.355818	joules (watt-sec)	J
	watt-hours	3600	joules (watt-sec)	J
	kilocalories (TC)	4184	joules (watt-sec)	J
	kilocalories (IT)	4186.8	joules (watt-sec)	J
	kilocalories (mean)	4190.02	joules (watt-sec)	J
	ergs (cm-dynes)	9.807×10^{-5}	joules (watt-sec)	J
	BTU (TC)	0.2929	watt-hours	Wh
	BTU (IT)	0.2931	watt-hours	Wh
	BTU (mean)	0.2933	watt-hours	Wh
	foot-pounds	3.766×10^{-4}	watt-hours	Wh
	kilocalories (TC)	1.1622	watt-hours	Wh
	kilocalories (IT)	1.1630	watt-hours	Wh
	kilocalories (mean)	1.1639	watt-hours	Wh
	ergs (cm-dynes)	2.7242×10^{-8}	watt-hours	Wh
ELECTRICITY				
Q	Coulombs	2.778×10^{-4}	ampere-hours	Ah
	Faradays	26.8	ampere-hours	Ah
	Faradays	96500	coulombs	
MAGNETIC FLUX				
phi	Webers	1×10^8	Maxwells (lines)	
gilbert		0.7958	amp-turn	
oersted		79.58	amp/meter	
MAGNETIC FLUX DENSITY				
	lines/sq. inch	0.155	Gausses (lines/cm^2)	
Wb/cm^2	Webers/sq. cm	1×10^8	Gausses (lines/cm^2)	
Wb/m^2	Webers/sq. meter	1×10^4	Gausses (lines/cm^2)	
Wb/in^2	Webers/sq. inch	1.55×10^7	Gausses (lines/cm^2)	
	lines/sq. inch	1.55×10^{-5}	Tesla	T
Wb/cm^2	Webers/sq. cm	1×10^4	Tesla	T
Wb/m^2	Webers/sq. meter	1	Tesla	T
Wb/in^2	Webers/sq. inch	1550	Tesla	T
Gauss		1×10^{-4}	Tesla	T
GRAVITATIONAL ACCELERATION				
		9.80665	meters/sec/sec	m/s^2
		32.17405	feet/sec/sec	ft/s^2
		21.93685	miles/hour/sec	mph/s
		35.30394	kilometer/hr/sec	km/h/s

Symbol	Given ⇒ Mult. by To Obtain ⇐ Divide by		to Obtain Given	Symbol
TORQUE DAMPING FACTOR				
in-lb/krpm	inch-pounds/kilo rev/min	0.001079	Newton-m/rad/sec	Nm/r/s
ft-lb/rpm		12.948	Newton-m/rad/sec	Nm/r/s
TORQUE SLOPE				
rpm/oz-in		14.83286	rad/sec/Newton-meter	r/s/Nm
MAGNETIZING FORCE				
Oe	oersteds	0.7958	Amp-turns/cm	
		2.021	Amp-turns/inch	
		79.58	Amp-turns/m	
MOTOR CONSTANT				
Ke	Volts/rpm	104.72	volt/rad/sec	

Mean calorie and BTU is used here throughout. The IT calorie, 1000 international steam-table calories, has been defined as the 1/860th part of the international kilowatthour.

Symbol	Given	Compute by:	To obtain	
TEMPERATURE				
°F	Farenheit	(°F − 32)/1.8	Centigrade	°C
K	Kelvin	1.8 K + 305.15	Farenheit	°F
°C	Centigrade	°C1.8 + 32	Farenheit	°F
K	Kelvin	+273.15	Centigrade	°C
°F	Farenheit	(°F−305.15)/1.8	Kelvin	K

INDEX

Electric Bicycles: A Guide to Design and Use, by William C. Morchin and Henry Oman
Copyright © 2006 The Institute of Electrical and Electronics Engineers, Inc.

187